Unfinished Symphony:
The Standard Model and the
Quest for a Complete Theory

Jacks

Table of Contents

Chapter 1: Introduction

1.1 Introduction

The Standard Model (SM) of particle physics is one of the best-tested theories in physics and accurately describes phenomena down to the smallest scales measured. The model describes a universe of fundamental particles and forces whose interactions determine the properties of the macroscopic world we see around us. With the experimental discovery of the Higgs boson (H) in 2012 [1, 2], last piece of the model was confirmed. It is however, known to be incomplete. Most notably, the SM includes only three of the four known fundamental forces, omitting gravity, and astrophysical observations tell us that the SM particles account for only 5% of the contents of the universe. Another 23% of the universe is dark matter, seen in the movement of galaxies, and 72% is dark energy, responsible for the accelerating expansion of the universe. Many extensions of the SM have been proposed, collectively called Beyond the Standard Model (BSM) theories, which for example, add new particles matching observed dark matter properties. However, despite the gaps in the model, observations contradicting the SM that could point to more complete theories are few and far between.

The Large Hadron Collider (LHC) is the largest particle accelerator ever built, designed to collide particles at energies that few natural processes can match. The unique environment created by the LHC allows for tests of the SM that cannot be performed anywhere else. The ATLAS detector, built and run by the ATLAS collaboration at CERN, is a general-purpose detector designed to study all types of SM particles, as well as many theorized new particles. In this book, a search is presented for two types of massive particles, a scalar (X) and tensor (G_{KK}^*), decaying to a pair of boosted Higgs bosons (HH). The Higgs bosons are further required to decay to pairs of b-quarks. The search is performed using 139 fb^{-1} of data collected from $\sqrt{s} = 13$ TeV proton-proton collisions using the ATLAS detector between 2015 and 2018.

This book is structured as follows:

Chapter 1 contains a brief introduction to high-energy physics, to particle detection technologies and to the ATLAS experiment. In Chapter 2, I present the search for heavy resonances decaying to $HH \rightarrow 4b$, detailing the analysis strategy and results. In Chapter 3, I describe the ongoing calibration of a new machine learning algorithm trained to identify particle decays to pairs of b-quarks. Finally, in Chapter 4, I discuss the results of this research as well as opportunities for future study.

1.2 High Energy Physics

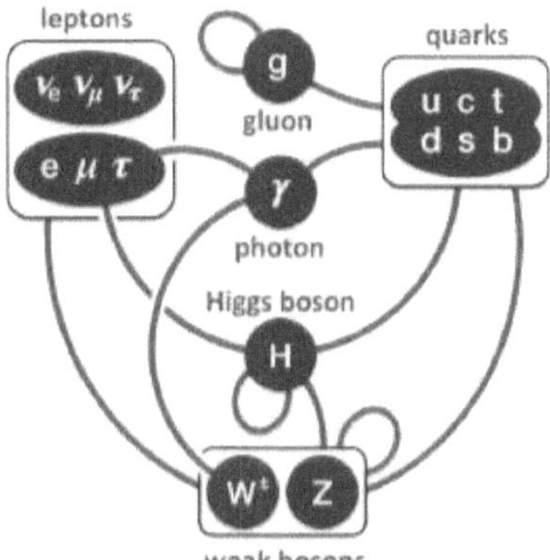

Figure 1.1: Diagram of the particle content of the Standard Model. Particles are arranged in groups with similar properties, and blue lines indicate which groups interact.

The Standard Model of particle physics describes the universe in terms of a set of elementary particles and the interactions between them. These particles have certain intrinsic properties, including mass, spin and charge, and manifest as indivisible excitations of quantum fields. In the SM, particles are classified by these intrinsic properties as shown in Figure 1.1. Matter is made up of spin-1/2 fermions, grouped into the quarks and leptons. Interactions between the matter particles

2

are mediated by spin-1 gauge bosons, each of which corresponds to one of the fundamental forces of nature. The electromagnetic force, for example, affects all particles with an electric charge (Q) and is mediated by the photon. The strong force is mediated by gluons and acts on particles with a color charge (r, g, b), and the weak force is mediated by the W and Z bosons and acts on weak isospin (T_3). In addition, each particle in the SM has a corresponding antiparticle that has the same mass and spin but opposite charge and parity.

The six quarks interact with all three of the electromagnetic, weak and strong forces. The quarks are grouped into three pairs consisting of an up-type quark (with $Q = +2/3$ and $T_3 = +1/2$) and a down-type quark (with $Q = -1/3$ and $T_3 = -1/2$). Due to the strong force, quarks are never found in isolation. They are always confined to composite particles, called hadrons, which exist only in states of net zero color charge or symmetric combinations of r, g, and b. The lightest quarks are the up (u) and down (d) and form the most commonly found hadrons, including protons, neutrons and pions. The up and down quarks have masses of $2.16^{+0.49}_{-0.26}$ MeV [1] and $4.67^{+0.48}_{-0.17}$ MeV respectively [3]. The second-generation quarks, the charm (c) and strange (s), have masses of 1.27 ± 0.02 GeV and 96^{+11}_{-5} MeV respectively [3]. The heaviest quarks are the bottom (b) and top (t), which have masses of $4.18^{+0.03}_{-0.02}$ GeV and 172.76 ± 0.30 GeV respectively [3]. Both are of particular importance to this book and will be discussed further in later sections.

The set of leptons are similarly organized into three pairs, with the electron (e), muon (μ), and tau (τ) having $Q = -1$ and $T_3 = -1/2$, and the three corresponding neutrinos (ν_e, ν_μ, ν_τ) having $Q = 0$, $T_3 = +1/2$. The electron, muon and tau have masses of 511 keV, 105.7 MeV and 1.78 GeV respectively while the neutrinos are all almost mass-less [3]. Unlike quarks, leptons carry no color charge and can be found in isolation.

Finally, the Higgs boson has no spin, no electric or color charge and a mass of 125.1 ± 0.14 GeV [3]. Unlike, the other bosons, it does not mediate a force but is instead a remnant of electroweak symmetry breaking, as explained in Section 1.2.2.

---

[1] In the units commonly used in high-energy physics, and throughout this book, the speed of light, c, and the reduced Planck constant, $\hbar$, are treated as dimensionless quantities with value one. As a consequence, the units of energy, electron-volts (eV), are used for momentum and mass, which properly have units of eV/c and eV/c^2 respectively.

1.2.1 Gauge Theory

The organization of the SM particles can be understood by considering the underlying symmetries of nature. As a relativistic theory, the SM Lagrangian ($\mathcal{L}_{\text{SM}}$) is invariant under global transformations of the Poincaré group: translations, rotations, and boosts. These are continuous symmetries of a homogenuous 3+1 dimensional spacetime. By Noether's Theorem, for each continuous global symmetry of the theory there must be an associated conservation law and conserved charge. For the Poincaré group, these are conservation of energy and momentum, angular momentum and of center of mass. Particles transform under rotations and boosts (collectively called Lorentz transformations) according to their spin, and can only appear in the Lagrangian in Lorentz-invariant combinations. The form in which a particle appears in the Lagrangian is used to determine its equations of motion through the principle of least action. Spin-0 scalars obey the Klein-Gorden equation (Eq. 1.1a), spin-1/2 spinors obey the Dirac equation (Eq. 1.1b) and spin-1 vectors obey the Proca equation (Eq. 1.1c).

$$(\partial^\mu \partial_\mu - m_\phi^2)\phi = 0 \tag{1.1a}$$

$$(i\gamma^\mu \partial_\mu - m_\psi)\psi = 0 \tag{1.1b}$$

$$\partial_\mu(\partial^\mu B^\nu - \partial^\nu B^\mu) + m_B^2 B^\nu = 0 \tag{1.1c}$$

The Lagrangian for scalar, spinor and vector fields, with no interactions between them, would look like:

$$\mathcal{L} = -\frac{1}{4}F_{\mu\nu}^a F_a^{\mu\nu} + i\bar{\psi}\gamma^\mu \partial_\mu \psi + \frac{1}{2}\partial_\mu \phi^\dagger \partial^\mu \phi + m_B^2 B^\nu B_\nu - m_\psi \bar{\psi}\psi - m_\phi^2 \phi^\dagger \phi, \tag{1.2}$$

where $F^{\mu\nu} \equiv \partial^\mu B^\nu - \partial^\nu B^\mu$ and the sign of the terms is set by convention. $\mathcal{L}_{\text{SM}}$ is more complicated, however, and particles interact according to sets of internal (gauge) symmetries.

In addition to the symmetries of spacetime, the SM Lagrangian is also invariant under several sets of internal symmetries corresponding to the fundamental forces. For example, the electromagnetic force is represented mathematically by a U(1) symmetry. Mathematically this corresponds to invariance under a local transformation by a complex phase, $\psi \rightarrow e^{i\alpha}\psi$ and $\psi^{\dagger} \rightarrow e^{-i\alpha}\psi^{\dagger}$. The spinor mass term, $m_{\psi}\bar{\psi}\psi$, is naturally invariant under this transformation but the kinetic term, $i\bar{\psi}\gamma^{\mu}\partial_{\mu}\psi$, is not. The gauge invariance of the kinetic term is restored by replacing ∂_{μ} with the covariant derivate $\mathcal{D}_{\mu} \equiv \partial_{\mu} + iqB_{\mu}$, where B_{μ} is a new vector field and q is the coupling constant of this field to the particle ψ. In this case, the new field corresponds to the photon and q to the electric charge. The U(1) symmetry of the electromagnetic force generates only a single vector field, but the method can be extended to more complex symmetry groups as well. Similar covariant derivatives are used to generate fields for the SU(3) gauge symmetry of the strong force and the SU(2)$_L$ symmetry of the weak force[2]. The full SM gauge group is $SU(3) \times SU(2)_L \times U(1)$ and generates three sets of vector fields: eight gluons, three weak bosons and one photon, respectively.

The full SM Lagrangian is then:

$$\mathcal{L}_{SM} = -\frac{1}{4}F_{\mu\nu}^a F_a^{\mu\nu} + i\bar{\psi}\gamma^{\mu}\mathcal{D}_{\mu}\psi - m_{\psi}\bar{\psi}\psi + \mathcal{L}_{\text{Higgs}}, \tag{1.3}$$

where $F_a^{\mu\nu} \equiv \partial^{\mu}B_a^{\nu} - \partial^{\nu}B_a^{\mu} + gf^{abc}B_b^{\mu}B_c^{\nu}$ is the field strength term for a gauge field with self-interactions. To keep the equation compact, all fields of the same type, i.e. spinor or vector, are represented by a single term, though the charge, g, the structure constant, f^{abc}, and the form of the covariant derivative, $\mathcal{D}_{\mu}$, are different for different fields. The mass term of the vector fields, previously written as $m_B^2 B^{\nu}B_{\nu}$, has been folded into the Higgs sector of the Lagrangian because it is not otherwise gauge invariant. The Higgs boson was first proposed as a mechanism to allow the observed W and Z boson masses to fit into the Lagrangian in a gauge-invariant way.

1.2.2 Higgs Mechanism

The Higgs Mechanism was proposed as an explanation for the large masses of the W and Z bosons. At high energies, the theory goes, the electromagnetic and weak forces are the same, governed by an SU(2)×U(1)$_Y$ symmetry and four massless gauge bosons. Importantly, since the

---

[2]The L subscript in SU(2)$_L$ indicates that the weak force acts only on left-handed chiral particles.

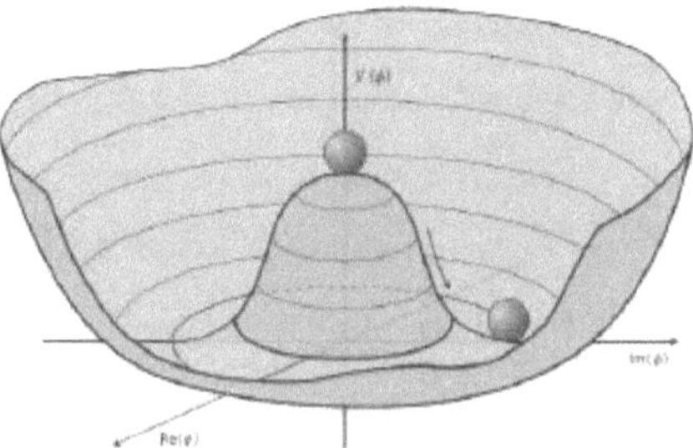

Figure 1.2: The 'Mexican-hat' potential of the Higgs field, V(ϕ) [4]. The symmetry of the potential is broken as the system falls into a stable ground state.

gauge bosons are massless, the Lagrangian is invariant under this electroweak symmetry. The electroweak theory also includes a complex scalar doublet, called the Higgs field, with hypercharge $Y = 1/2$. This field has an oddly shaped potential, shown in Figure 1.2 and often referred to as the 'Mexican-hat' potential. Importantly, the lowest energy state of this potential has a non-zero field strength, whose value is called the vacuum expectation value or vev. Furthermore, the field has not just one lowest energy state but a continuous set of them characterized by a complex phase. These states are functionally identical but a cold universe can only exist in one of them, thus spontaneously breaking the symmetry.

Typically, the Higgs potential is written in the simplest form that provides a non-zero vev:

$$V(\phi) = \frac{1}{2}\partial_\mu\phi\partial^\mu\phi - m_\phi^2\phi^2 + \lambda\phi^4. \tag{1.4}$$

More generally though, any Higgs potential with a non-zero vev would produce spontaneous symmetry breaking and studying the shape of the Higgs potential is a major goal of high-energy physics. After symmetry breaking, the Higgs potential can be rearranged to leave a single massive particle, h:

$$\phi(x) = \sqrt{\frac{1}{2}}\begin{pmatrix} 0 \\ v + h(x) \end{pmatrix}. \tag{1.5}$$

In the so-called Higgs gauge, the electroweak gauge bosons are reorganized into the more familiar photon, $W^{\pm}$ and Z. Additional mass terms for the Higgs field and the weak bosons appear in the Lagrangian in gauge-invariant combinations, and in this sense the Higgs 'generates' the mass of the W and Z bosons. The original Higgs mechanism has since been extended to give all elementary particles mass. Yukawa interactions between fermions and the Higgs field are a gauge-invariant way to add mass-like terms as seen for the bosons. The observed particle mass is then interpreted as a measure of the interaction strength between that particle and the Higgs boson. The Higgs boson was first discovered by the ATLAS and CMS collaborations in 2012, and thus far measurements of its properties match closely with SM predictions. However, the Higgs boson presents several unique challenges to the SM and many hope that its study will shine light on BSM physics.

1.3 Beyond the Standard Model

The Standard Model of particle physics is known to be incomplete, but there are many different ways in which it can be expanded. Gravity, if it were to be included in the theory, would be mediated by a mass-less spin-2 boson called the graviton. The force of gravity is, however, extraordinarily weak relative to the other forces of the SM and the effects of gravity on particle interactions are unclear. The fact that gravity is a factor of $O(10^{16})$ smaller than the other forces precludes it from being added to the SM, and is called the "hierarchy problem". The observed mass of the Higgs boson also introduces the so-called "naturalness problem". Due to its scalar nature, the observed mass of the Higgs boson is affected by radiative corrections from e.g. the loop diagrams in Figure 1.3. In theory, these corrections should push the observed mass up based on the highest mass particles it interacts with. Since the Higgs boson interacts with all massive particles, this observed mass should be at the highest energy scale in physics, that of gravity. In the SM this can be explained away by saying that the 'bare' mass of the Higgs boson, before corrections, cancels the corrective term out to 16 digits. Many argue that a coincidental cancellation of that magnitude would be unnatural and must be explained by some addition to the theory. There are many models of BSM physics that can solve these theoretical difficulties, two of which are briefly introduced below. The Randall-Sundrum (RS) model [5, 6, 7, 8] and the two-Higgs doublet model

(2HDM) [9, 10, 11] are used as "benchmark" models in the $HH \rightarrow 4b$ search presented in this book. Both introduce some additional physics to the Higgs sector of the Lagrangian, and both predict striking testable predictions for heavy resonant di-Higgs production at the LHC.

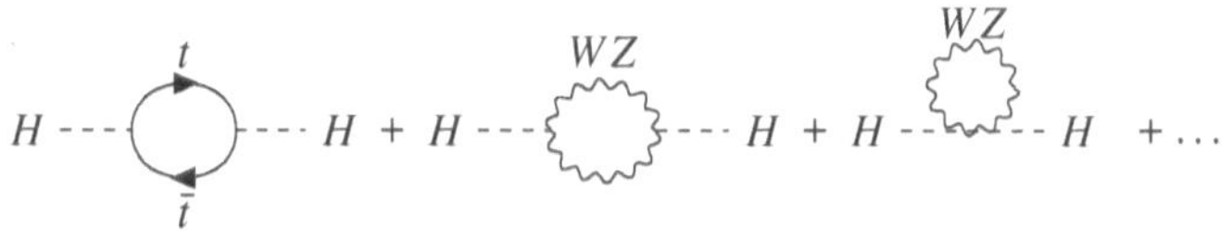

$$H \text{---} \bigcirc \text{---} H \; + \; H \text{---} \{WZ\} \text{---} H \; + \; H \text{---} WZ \text{---} H \; + \ldots$$

Figure 1.3: Example loop diagrams contributing NLO corrections to the Higgs boson mass. No mechanism exists in the SM to cancel these large radiative corrections.

1.3.1 Randall-Sundrum Model

The Randall-Sundrum warped extra dimension model, first proposed by Lisa Randall and Raman Sundrum in 1999, is characterised by the existence of an additional finite spatial dimension. The metric of this spacetime contains a "warp" factor applied to the traditional four-dimensional metric that varies exponentially along the additional dimension:

$$ds^2 = e^{-2kr_c\phi}dx^\mu dx_\mu + r_c^2 d\phi^2, \tag{1.6}$$

where x^μ are coordinates of the familiar spacetime dimensions, $\phi \in [0, \pi]$ is the coordinate of an extra dimension, while k and r_c are free parameters of the model. The exponential warping of 4-dimensional spacetime along the additional dimension can create large hierarchies in scale with modest values of the dimensionless combinations kr_c and $k/\overline{M}_{Pl}$, where $\overline{M}_{Pl} = 2.4 \times 10^{18}$ GeV is the effective four-dimensional Planck scale.

The RS model predicts a distinctive set of new particles. The existence of a finite extra dimension necessitates a set of resonant modes, called Kaluza-Klein modes, visible to a four-dimensional observer. The four-dimensional observer does not see the momentum of a particle along the extra dimension, instead interpreting this additional energy as rest mass. Since the extra dimension is finite, it must have a discrete set of momentum states (consider e.g. the harmonics of a string) that the four-dimensional observer will see as massive resonances. Furthermore, the masses of these resonances depend on the size and shape of the additional dimension. A tower of Kaluza-Klein resonances would therefore be both strong evidence for the RS model and allow for measurements of the model parameters. In the $HH \rightarrow 4b$ analysis, only the decay of the lowest-mass Kaluza-Klein graviton state into a pair of SM Higgs bosons is considered, as shown in Figure 1.4a.

1.3.2 Two-Higgs Doublet Model

The 2HDM is the simplest extension of the Higgs sector of the SM [9]. While it is not a UV-complete model by itself, constructions like the 2HDM are components of many BSM theories, such as the Minimal Supersymmetric Standard Model (MSSM) [10, 11]. In essence, the Higgs field ϕ described in Section 1.2.2 is replaced by a pair of fields, ϕ_1 and ϕ_2, related by a U(2) symmetry. With two scalar doublets, the Higgs sector in the 2HDM has eight degrees of freedom rather than four and, after electroweak symmetry breaking, predicts five massive bosons instead of one. In the 2HDM, the observed Higgs boson is associated with the lighter of two neutral, CP-even bosons, H and X. The model also predicts a pair of electrically-charged Higgs bosons ($H^{\pm}$) and a neutral axial boson (A). The $HH \rightarrow 4b$ analysis considers decays of the neutral heavy state X into a pair of the lighter Higgs states H, as shown in Figure 1.4b.

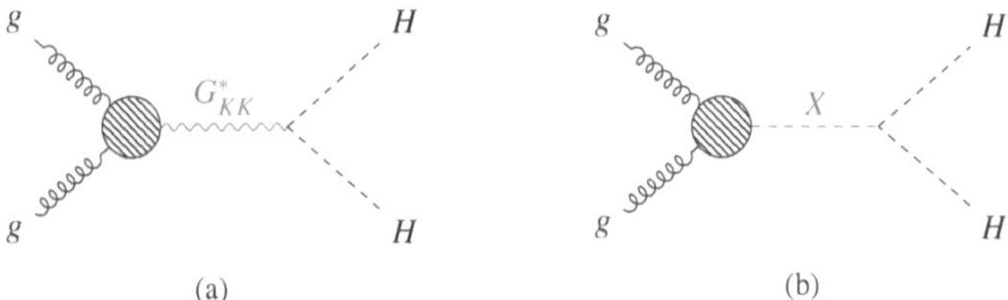

Figure 1.4: Feynman diagrams for the (a) RS graviton and (b) 2HDM scalar signal models used in the $HH \rightarrow 4b$ analysis.

1.4 The Large Hadron Collider

The LHC [12] is the largest and highest energy particle accelerator in the world. The LHC is a circular accelerator with a 27 km circumference, built 100 m under the French-Swiss border near Geneva, Switzerland. During operation, the LHC circulates two beams of protons in opposite directions around the ring, and crosses the beams at designated collision points. Detectors are built around each of the collision points to measure particles produced in these high-energy collisions. A diagram of the main LHC ring, along with the detectors and pre-accelator ring, is shown in Figure 1.8.

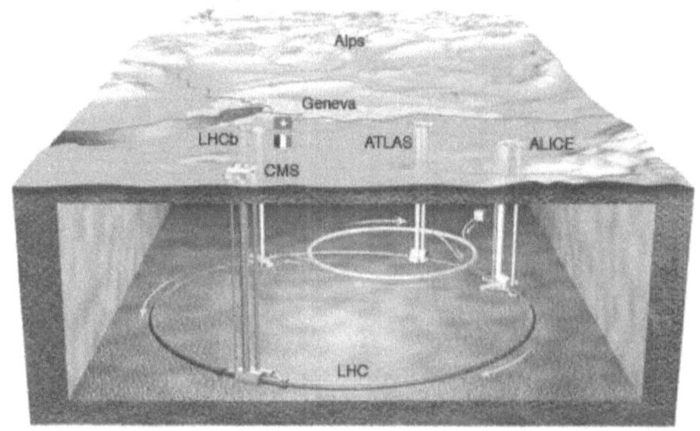

Figure 1.5: Diagram of the LHC tunnel and the detectors built around collision points [13].

The LHC uses magnetic fields created by superconducting magnets to control the beams. These beams are not continuous, but instead consist of discrete bunches of up to 10^{11} protons. Each bunch is accelerated through a series of pre-accelerators to an energy of 450 GeV before being injected into the main LHC ring. The protons are then accelerated up to a final energy before crossing the beams at the interaction points. Proton bunches are spaced around the ring so that collisions, or bunch crossings, occur at 25 ns intervals. LHC operation between 2015 and 2018 is collectively called Run 2. During this time up to 2556 bunches of protons were injected into the LHC ring at a time and accelerated up to 6.5 TeV. The LHC reached a record peak luminosity of $2 \times 10^{34} \text{cm}^{-2}\text{s}^{-1}$

in 2018 and delivered a total integrated luminosity of $160\,\mathrm{fb}^{-1}$ in Run 2.

1.5 Collider Physics

Luminosity is a measure of the rate of collisions produced by the experiment, defined as the number of particle interactions per unit area and time. To be precise, luminosity is defined as:

$$L = \frac{nN_1 N_2 f}{4\pi \sigma_x \sigma_y},$$
(1.7)

where n is the number of particle bunches in the beams, N_1 and N_2 are the number of particles per bunch in each beam, f is the bunch crossing frequency, and σ_x and σ_y describe the width of the bunches in the plane transverse to the motion. Knowing the luminosity and the cross-section for a specific process allows one to predict how many times that process should occur within a given timeframe. For a process $pp \rightarrow X$ with cross-section $\sigma(pp \rightarrow X)$, the expected number of times it occurs is

$$N(pp \rightarrow X) = \int L\sigma(pp \rightarrow X)dt.$$
(1.8)

The center-of-mass energy of LHC collisions, $\sqrt{s}$, is the combined energy of the proton beams, $13\,\mathrm{TeV}$ [14]. However protons are composite particles made of quarks, anti-quarks and gluons, collectively called partons. When the LHC collides bunches of protons, interactions occur between pairs of the partons they contain, each of which contains only a fraction of the proton's energy. The energy fraction carried by each type of parton is described by a set of parton distribution functions (PDFs) calculated primarly from Deep Inelastic Scattering experiments.

The goal of the LHC is to study hard scatter processes, i.e. parton-parton interactions involving large momentum transfers. Figure 1.6 shows a schematic diagram of a proton-proton collision. The initial hard-scatter process can be represented by Feynmann diagrams which are often exactly calculable to leading-order (LO) or next-to-leading-order (NLO). The propagation of any quantum chromodynamics (QCD) remnants, i.e. any non-color-singlet particles in the final state, is modelled by the DGLAP [16, 17] evolution equations up to the point at which they form color-singlet bound

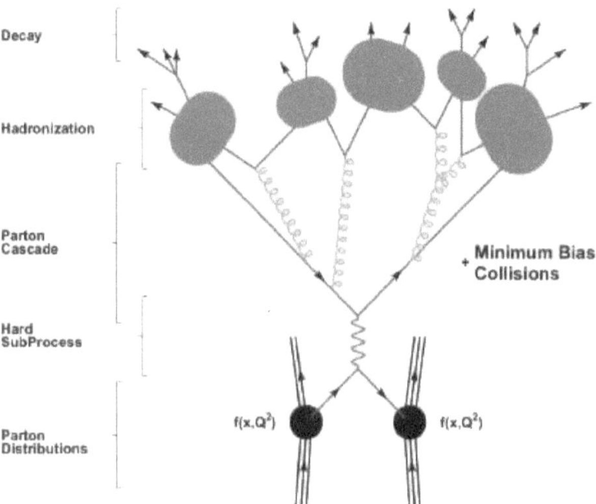

Figure 1.6: Schematic diagram showing the stages of a simulated hadron collision [15]. Initial energies of the incoming partons are set by parton distribution functions. The hard scatter process commonly produces 'bare' quarks which shower, creating a complex set of quarks and gluons, before combining into color-singlet states in a process called hadronization. Finally, the hadrons decay into stable states.

states in a process called hadronization. The parton shower is brief, completing as color-charged particles travel past the femtometer-scale distance at which QCD is weakly interacting. In the brief period before hadronization, high-energy particles radiate large numbers of gluons and light quarks, resulting in collimated sprays of hadrons, called jets, from a single source. Finally, these hadrons decay into stable states which propagate outward from the collision point to be measured in the particle detectors. There are many sources of low energy radiation that occur simulateously with the hard scatter process, including the radiation of quarks and gluons from the initial state partons (ISR) or from the final state (FSR), collisions of other partons in the same protons or of other protons in the bunch. These other collisions are collectively called pile-up and represent an important source of background radiation in the detector. The average number of collisions per bunch crossing was $\langle\mu\rangle = 13.4$ in 2015, increasing up to $\langle\mu\rangle = 36.1$ in 2018. Higher luminosities produce more data but also create more pile-up interactions in the detector.

1.6 Common Detector Technologies

There are many different technologies used to measure the energies, momenta and trajectories of particles. This section provides a general overview of several common classes of detectors used in ATLAS. Broadly speaking, detectors consist of an active component that reacts with passing particles, a read-out system to record those interactions, and often a passive structural component. There are many ways particles can interact with detector materials and the details are critical to detector design and construction. Here though, I will only discuss generalities. Ignoring the details of specific processes, electromagnetic interactions can come in the form of an electron radiating photons (e.g. 'bremsstrahlung'), a photon splitting into an electron-positron pair, or a transfer of energy to the valence electrons of an atom (often stripping them from that atom). Hadronic interactions with atomic nuclei are much rarer but involve larger transfers of energy. Detectors built to measure nuclear interactions tend to be larger and denser than electromagnetic detectors. Some particles, such as neutrinos, interact so rarely that they can escape the ATLAS detector entirely. Experiments designed to detect these interactions can require tons of active material and/or be hundreds of meters across. Electromagnetic and hadronic interactions with the detector material transfer energy and can be measured in several different ways, depending on the material and the type of interaction. This section will focus on two types of detectors, ionization chambers and scintillators, as well as a brief introduction to calorimeters, detectors that measure particle energies.

1.6.1 Ionization Detectors

Ionization detectors are built to generate electrical signals from free electrons in the detector material. An ionization detector can be quite simple, consisting of a gas held in a uniform electric field between two metal plates. Atoms in the gas are normally electrically neutral, but charged particles passing through the gas leave trails of ions and free electrons. These electrons and ions are pulled to opposite ends of the detector by the electric field and induce a current on the plates as they move. The size and shape of the induced current pulse depends on the shape of the detector and the strength of the electric field applied, as well as the intrinsic speed of the electron and ion in the gas. By tuning the electric field, one can control the acceleration of the ionized particles and the size of the detector response. Figure 1.7 shows how the current induced by an incident particle changes with applied electric field. In the ionization chamber region, ions are created only by the passage of incident particles through the gas. At high field strengths, however, the electrons accelerate enough to ionize other atoms in the gas. These are called secondary ionizations and create cascades near the anode of the detector. In the proportional counter region, secondary ionizations amplify the signal but the total current is still proportional to the number of primary ionizations. Finally, some detectors operate in the Geiger-Müller region, where each incident particle saturates the gas with cascading ionizations until the detector is reset.

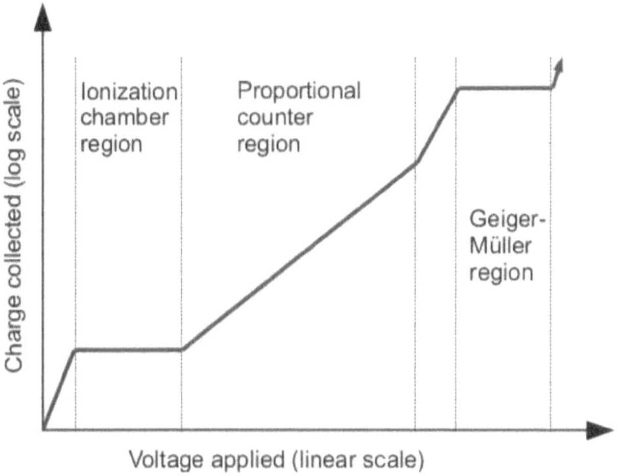

Figure 1.7: Simplified diagram of ionization detector response as a function of applied voltage.

An important consideration when designing any detector is the recovery time, that is, the time it takes for the detector to return to a neutral state. Ideally, the full signal induced by one particle should be collected before another passes through. The ATLAS collaboration uses solid silicon sensors for ionization detectors with very short recovery time. These silicon sensors operate on a similar principle to the gas drift chamber described above, but have a higher number of primary ionizations per unit length, higher drift velocity, higher radiation tolerance and higher cost. Each silicon sensor has a layer of p-type silicon (enriched in positive charges) deposited on an n-type silicon substrate (enriched in negative charges). At the boundary, charges from one layer combine with the other, forming a charge-depleted zone in the center that functions as the detector. Most electrons in the silicon are initially in a set of low energy bound states, called the valence band. Charged particles passing through the depletion zone excite electrons into the higher energy conduction band states. Exciting an electron to the conduction band allows it to travel freely through the material, and creates a hole in the valence band that can travel like a positively-charged electron. An applied electric field causes the conduction electron and the hole to drift apart, inducing current on the anode and cathode of the detector in a similar manner to the electron-ion pairs in a gas detector. The higher density of a solid detector means that it will have more interactions per unit length, and can remain sensitive while being much smaller. These sensors are used when recovery time is critical, e.g. in the detector components nearest the LHC beamline, which must fully reset in the 25 ns between bunch crossings.

1.6.2 Scintillation Detectors

Scintillation detectors generate signals from light rather than from free electrons. When a particle passes through a material it can excite electrons into higher-energy bound states. These excited states are unstable and the electron will eventually radiate that energy as a photon and drop back to the ground state. In most materials, a photon produced this way would be reabsorbed to excite a nearby atom, but for a scintillator to be effective the light needs to be able to travel uninterrupted to the edges of the detector. This means that the passage of an energetic particle must put the atom into an energy level that the bulk of the material cannot readily access. In organic scintillators, generally, the decay of the excited state goes through an intermediate metastable state, while inorganic crystals are typically doped with low concentrations of impurities to create energy levels that the pure material cannot access. In either case, the scintillation light travels to the edges of the detector where it is converted to an electrical signal by e.g. a photomultiplier tube. Unlike an ionization detector, where recovery time depends largely on the drift of ions to the anode and cathode, scintillator recovery time depends on the decay rate of the excited states. Recovery time, as well as the wavelength and yield of photons, vary significantly from one material to the next. Scintillators are also often used in calorimetery, where denser materials and larger components with more stopping power may be required.

1.6.3   Tracking and Calorimetry

The ATLAS subsystems can be divided into two categories: tracking detectors and calorimeters. These distinctions are separate from the type of material interaction used, and instead depend on the purpose of the detector. The tracking detectors are built to optimize positional resolution and measure the direction of a particle trajectory. The calorimeters are optimized to measure particle energies, generally absorbing the incoming particles in the process. A key consideration when designing either is the radiation length of a material, X_0, which is approximately the thickness of material necessary to reduce a particles' energy by a factor of e. Interactions in which a high-energy particle loses a significant amount of energy generate showers of particles, each with enough energy to create further detector signals. These showers are typical of calorimeters, which are built many radiation lengths deep. Often calorimeters are built as sampling detectors, where an active material is layered with dense passive absorbers. These absorbers are often structural elements, but also initiate particle showers through bremmstrahlung or nuclear interactions. A downside of the sampling technique is that particles can be trapped in the absorbing layers without contributing to the energy measured by the calorimeter. Precisely characterizing the fraction of energy deposited in the absorbers is a critical part of constructing a sampling calorimeter. Tracking detectors, by contrast, are typically built to minimize passive material and allow particles to pass through without losing much energy. Tracking detectors and calorimeters provide complimentary information and ATLAS combines both, built with a variety of active and passive components to optimize detection efficiency.

1.7   The ATLAS Detector

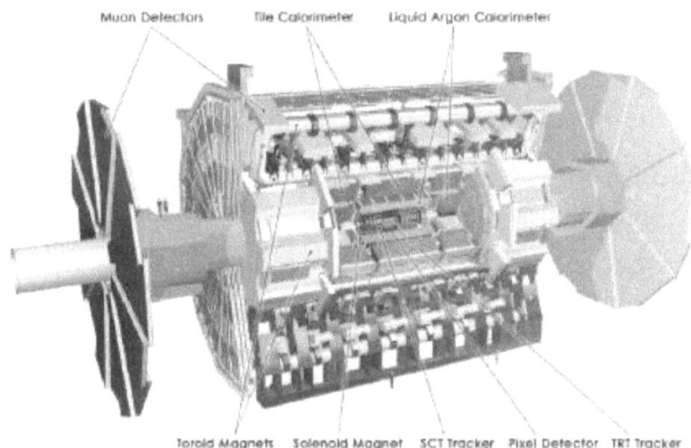

Figure 1.8: Schematic diagram of the ATLAS detector and its components. The detector and the people in the diagram are shown to scale. [18]

The ATLAS detector (**A** **T**oroidal **LHC** **A**pparatu**S**) [19] is a general-purpose particle detector located at one of the four interaction points of the LHC. The largest of the LHC detectors, it is 46 meters long, 25 meters tall and weighs 7,000 tons, almost as much as the Eiffel Tower. ATLAS is built in cylindrical layers, with sub-detectors positioned concentrically around the beamline, as shown in Fig. 1.8. The innermost layers, called the inner detector (ID), are tracking detectors intended to measure the track of charged particles without affecting their trajectory. Surrounding the ID are two layers of calorimeters, which can catch most particle species and measure the energy that they release. Finally, an outer layer of detectors, called the muon system (MS), provides precision tracking for muons, which are able to pass through the calorimeters without being absorbed. Most layers are further segmented into 'barrel' components in the center and 'endcaps' on the ends. The ATLAS subsystems each provide different and complementary information that can be combined to uniquely identify most particles passing through them.

Integral to the detector measurements are two sets of strong magnetic fields. A solenoid placed between the ID and the calorimeters creates a 2 T field parallel to the beam in the ID. The toroidal magnet for which ATLAS is named creates a field in the MS of approximately 0.5 T in the central region and 1 T in the endcaps. These fields bend the trajectory of charged particles in predictable ways, allowing the tracking detectors to distinguish between positive and negative charges and improving the resolution of momentum measurments.

1.7.1 Detector Geometry and Coordinate System

The ATLAS collaboration uses a right-handed coordinate system defined by the LHC beamline, which travels lengthwise through the detector and is designated as the z-axis. The x-axis points inwards, towards the center of the LHC ring, and the y-axis points upwards. Vector quantities are generally described in a modified cylindrical coordinate system, by the magnitude transverse to the beamline, the angle in the transverse plane, ϕ, and the pseudo-rapidity, η. Pseudo-rapidity is a function of the azimuthal angle, $\eta = -\ln(\tan\theta)$, that is equivalent to rapidity for massless, or highly relativistic, particles. (Pseudo-)rapidity is useful in high-energy physics as a measure of velocity that transforms additively under boosts, as opposed to velocity itself which transforms in a more complicated way. The transverse momentum, p_T, is particularly useful for collider physics as momentum conservation dictates that the p_T of all particles from one collision must be zero. This can be used to indirectly measure radiation that the detector cannot otherwise see, such as neutrinos.

The geometry of the detector is optimized to measure particles travelling outward from the collision point in the center. Most ATLAS subsystems are divided into three segments, a barrel and two endcaps. The barrel covers the central region of the detector and these components are generally mounted in concentric cylinders parallel to the beamline. The barrel is intended for high-precision physics and measures particles with large deflections from the beam, i.e. high p_T. The endcaps measure as much as possible of the particles that escape the barrel and are mounted perpendicular to the beamline. While the coverage of the barrel varies between subsystems, the overall detector is designed to optimize performance in the 'central' region of $|\eta| < 2.5$.

1.7.2 Inner Detector

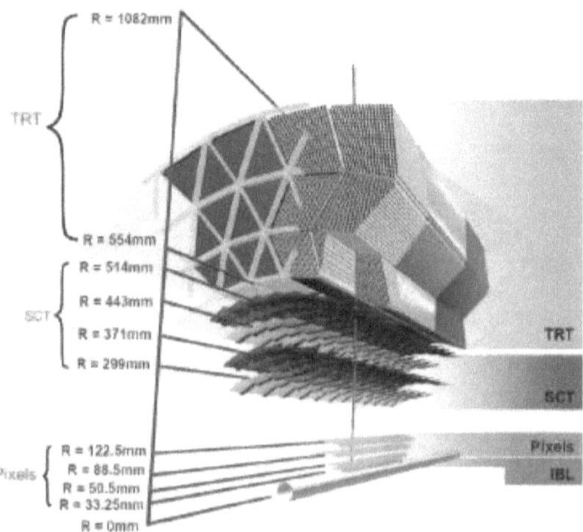

Figure 1.9: Diagram of the ATLAS inner detector systems [20]. The inner detector provides high-resolution tracking of charged particles coming from LHC collisions.

The ATLAS inner detector is composed of three subsystems: the pixel layers, including the insertable b-layer (IBL), the semiconductor tracker (SCT) and the transition radiation tracker (TRT) [20, 21]. In the barrel these subsystems are arround in concentric layers as shown in Figure 1.9. The pixel detectors are made of 50 μm×300 μm silicon pixels oriented so that particles bending in the magnetic field will traverse the long edge of the pixel. There are four layers of pixel detectors in the barrel with the innermost, the IBL, sitting 3 cm from the beamline. The pixel detectors have the best spatial resolution of any of the ATLAS subsystems as well as high radiation tolerance. The high precision of these innermost layers is critically important for the b-tagging algorithms used in this book. The two innermost layers provide coverage out to $|\eta| < 2.5$, while the other two cover out to $|\eta| < 1.7$. Four endcap disks of the same technology provide additional coverage in the $|\eta| = 1.7$-2.5 region.

Moving outward from the pixels, four layers of silicon strip detectors form the SCT. The silicon strip detectors are 6.36×6.4 cm^2, much larger than pixels, and are grouped into modules of four detectors each. The detectors are glued together at a small angle to obtain better precision in the z direction. The barrel layers provide coverage out to $|\eta| < 1.4$ with an additional nine encap wheels covering the $|\eta| = 1.4$-2.5 region.

The TRT consists of gas-filled straw tubes, 4 mm in diameter, interspered with layers of thin polypropelene foil. The straw tubes operate as drift chambers. In addition, particles entering or exiting the foil emit transition radiation photons, typically $O(10\,\mathrm{keV})$, which are detected by the straws, amplifying particle signals. The TRT covers the radial range of 56-107 cm from the beamline, and is arranged into barrel and endcap portions so that any particle will pass through approximately 36 tubes.

Each of the subsystems of inner detector produce "hits" when charged particles pass through them. The set of hits for each bunch crossing is reconstructed into a set of "tracks", representing the paths of particles moving through the detector.

1.7.3   Calorimeters

ATLAS has two calorimeter systems, the liquid argon (LAr) calorimeter [22] and the tile calorimeter (TileCal) [23]. The LAr calorimeter is a sampling calorimeter with the liquid argon interspersed with accordion-shaped lead-stainless-steel electrodes. The accordion plates are self-supporting and ensure material density is uniform as a function of azimuthal angle. Figure 1.10a shows the cross-section from a segment of the LAr barrel, with the accordion plates running through four layers of LAr cells. The LAr barrel covers a radial range of 1.5-1.97 m and $|\eta| < 1.4$, with endcaps providing coverage out to $|\eta| < 2.5$. Both barrel and endcaps are designed for electromagnetic calorimetry, that is, to measure electrons and photons as well as the light mesons that decay electromagnetically. The total radiation thickness to electrons and photons varies with η, but is at least $24X_0$ everywhere.

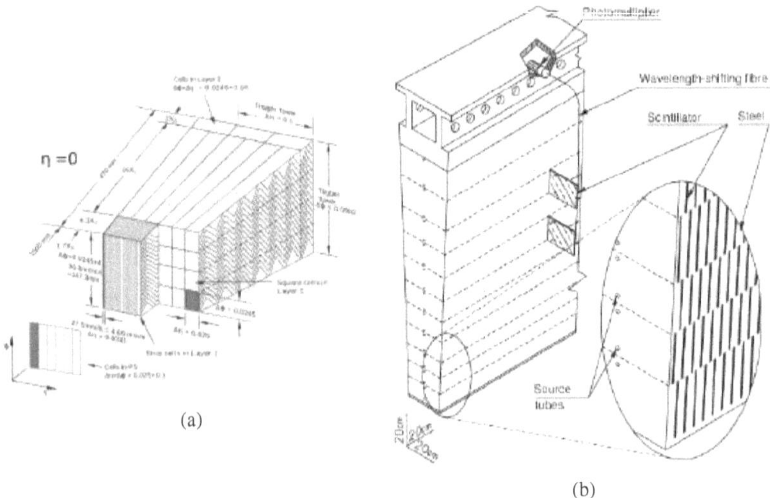

Figure 1.10: Diagrams showing the internal structure of the ATLAS (a) LAr and (b) Tile calorimeter systems [22, 23]. Liquid Argon cells in the LAr calorimeter are arranged into concentric layers and supported by accordian-shaped lead-stainless-steel structures. The plastic scintillating tiles used in the Tile calorimeter are supported by a steel structure, which also holds the photomultiplier tubes used to convert scintillation light to an electric signal.

The energy resolution of a calorimeter is parametrized with three terms,

$$\frac{\sigma_E}{E} = \frac{a}{\sqrt{E}} \oplus \frac{b}{E} \oplus c, \tag{1.9}$$

where a is the sampling term, b is the noise term and c is a constant term. The LAr calorimeter resolution for electrons was measured to be,

$$\frac{\sigma_E}{E} = \frac{10\%}{\sqrt{E}} \oplus \frac{170\text{MeV}}{E} \oplus 0.7\%. \tag{1.10}$$

Heavy hadrons lose little energy to electromagnetic interactions and, to these particles, the electromagnetic calorimeter is only a few X_0 thick. Separate detectors are therefore used for hadronic calorimetery, although these are still LAr calorimeters outside the barrel. The hadronic endcap calorimeters (HEC) are built from LAr and copper plates, and cover a range of approximately $|\eta| =$ 1.4-3.2, and additional forward calorimeters (FCAL) provide coverage out to $|\eta| < 4.9$. The FCAL is designed to be particularly dense due to the high level of radiation in the forward region. It is built in two sections, each comprised of a metal matrix with regularly spaced tubes housing metal rods and LAr filling the small gaps between rod and tube. The first section is built from copper while the other is tungsten.

For hadronic calorimetry in the central region, ATLAS uses a sampling calorimeter built from steel and plastic scintillating tiles. The scintillating tiles are read out through wavelength-shifting fibers attached on the ends, which carry the scintillation light to photo-multiplier tubes. The tiles are arranged in layers, as shown in Figure 1.10b, with fibers running along the outside. TileCal is designed to ensure a total thickness of at least $11X_0$ to hadrons, which lose energy primarily to nuclear interactions. The material in front of the TileCal is around $3\text{-}4X_0$ thick, mostly coming from the LAr calorimeter. The energy lost before a jet reaches the detector is called noncompensation and, for TileCal, causes a significant reduction of the precision of the energy measurement. The TileCal resolution for pions was measured to be,

$$\frac{\sigma_E}{E} = \frac{50\%}{\sqrt{E}} \oplus 6\%, \tag{1.11}$$

with a negligible contribution from electronic noise. The constant term is dominated by noncompensation.

1.7.4 Muon Spectrometer

The muon spectrometer [24] is the outermost set of ATLAS subsystems. These detectors are designed to cover the $5500\,\mathrm{m^2}$ surface area of the detector at a fraction of the cost of silicon pixels. The muon systems are gaseous ionization detectors used for additional measurements of muons from LHC collisions, and to detect cosmic rays entering the detector from above. The muon spectrometer is composed of three layers in both the barrel (out to $|\eta| < 1.0$) and endcaps ($|\eta| = 1.0$-

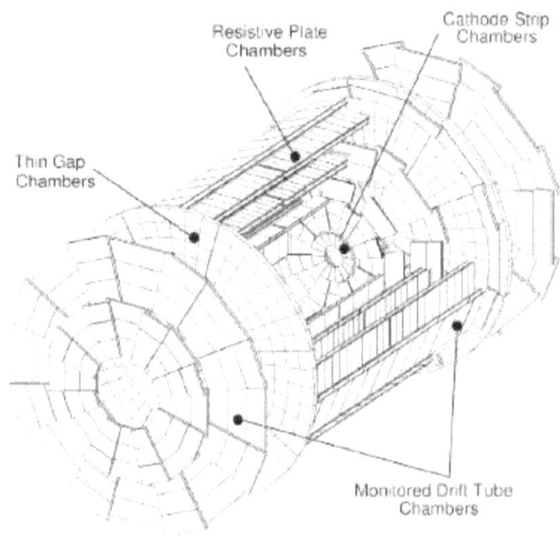

Figure 1.11: Schematic diagram of the ATLAS muon system [24].

2.7). It is surrounded by superconducting toroidal magnets the bend muon trajectories for better momentum resolution. The muon spectrometer is composed primarily of aluminum monitored drift tubes (MDTs), arranged in the bending direction of the magnetic fields. The tubes are 3 mm in diameter, filled with nonflammable gas, and have a W-Re wire running through the middle. The wire is held at 3.3 kV to generate an electric field, and the tubes operate in the same manner as the parallel-plate detector described in Section 1.6.1. In the innermost ring of the endcaps, cathode strip chambers (CSCs) are used instead. These are multi-wire chambers with a field generated by a series of wires spaced 2.54 mm apart, placed 2.54 mm from the readout cathodes. Measuring the current induced on multiple wires improves the position resolution of each chamber and reduces drift time. Additional information is provided by three layers of resistive plate chambers (RPCs) in the barrel and thin gap chambers (TGCs) in the endcaps. The RPCs are simple parallel-plate chambers while the TGCs are multi-wire chambers, both are operated at higher voltages than the

MDTs and CSCs for increased sensitivity and time resolution at the cost of momentum resolution. The RPCs and TGCs are used to identify the bunch crossing a muon came from as well as to provide coarse tracking and momentum measurements for the ATLAS trigger system.

1.7.5 Trigger

The high luminosity provided by the LHC represents a significant challenge for the ATLAS detector. While the collider is running, bunch crossings occur at a rate of 40 MHz with each bunch crossing generating approximately 1.6 MB of data. This is far more data than ATLAS can afford to store, and furthermore, the vast majority is of little interest. The total cross-section for inelastic proton-proton scatter is $O(100\,\text{mb})$ [3] while the cross-section for e.g. Higgs boson production is much smaller, $O(50\,\text{pb})$ [25]. Therefore, an automated procedure, the trigger, is used to quickly determine whether the data from any given bunch crossing should be recorded or simply discarded. The trigger consists of two steps called the Level-1 (L1) trigger and High-level trigger (HLT), which together reduce the event rate from the initial 40 MHz to only 100 Hz [26].

The L1 trigger is implemented entirely in hardware and makes a decision based on simplified trigger objects reconstructed from calorimeters and muon detectors. The trigger decision takes about 2.5 μs to calculate and keeps events with e.g. high p_T objects or large p_T imbalances across the detectors. The trigger thresholds are adjusted depending on luminosity to maintain an overall rate of 100 kHz sent to the HLT. If the L1 trigger is fired, then the event is retrieved from hardware storage buffers and transfered to the HLT CPU farm. The L1 trigger also identifies a set of regions of interest (RoIs), areas of the detector which recorded potentially interesting objects. The HLT is a pair of software-based triggers, performing a more detailed analysis of the event and further reducing the event rate to 100 Hz. While the reconstruction used at L1 is simplistic, generally simply counting the total energy recorded in an area of the detector, HLT uses the same techniques and object definitions to those described in Section 1.9. As a first step HLT reconstructs objects only within the identified RoIs, this step typically uses only $\sim$ 2% of the full detector information, takes $O(10\,\text{ms})$, and reduces the event rate to 1 kHz. Events passing this step are fully reconstructed, which takes $O(1\,\text{s})$ per event, for the final trigger decision. The criteria HLT uses for keeping an event are often similar to those of L1, but the multi-stage trigger design allows for more detailed requirements, including e.g. b-tagging similar to that discussed in Section 3.2. The triggers used in this book merely require events to contain a large, high p_T jet in the central region of the detector.

1.8 Particle Identification

Event reconstruction is the process of transforming the myriad detector signals generated by a signal bunch crossing into a useful picture of the hard scatter process. The basic building-blocks of event reconstruction are tracks, vertices and 'topo-clusters'. Both tracks and vertices are reconstructed primarily using the ID. Since charged particles radiate outward from the collision along predictable spiral trajectories one can, in essence, connect the dots to turn a series of pixel hits into a track. Of course, this becomes much more complex in a dense environment with many particles [27]. Track reconstruction is the most computational expensive part of ATLAS event reconstruction and is expected to become more difficult with planned machine upgrades. Collision vertices are identified from points on the beamline where large numbers of tracks intersect. For most purposes, including the analyses presented in this book, the primary vertex is identified as the one with largest H_T, defined as the scalar sum of the p_T of all tracks from that vertex [28]. Calorime-ter signals are grouped into sets of topologically-connected cells called 'topo-clusters' [29]. A topo-cluster does not necessarily represent the full shower caused by a single particle, but could correspond to a shower fragment or a cluster due to several overlapping showers. Topo-clusters are built around seeds, cells with measured energy four times greater than a background noise threshold, by adding neighboring cells based on their signal to noise ratios. A procedure is then applied to merge overlapping topo-clusters, and to split those clusters with distinct local maxima. Finally, the resulting set of topo-clusters are calibrated to correct for differences between electromagnetic and hadronic responses, signal losses due to the clustering algorithm, and energy lost in the passive materials. Fig. 1.12 shows an example event reconstruction, with tracks radiating from the collision vertices drawn as colored lines through the ID and topo-clusters shown as boxes radiating

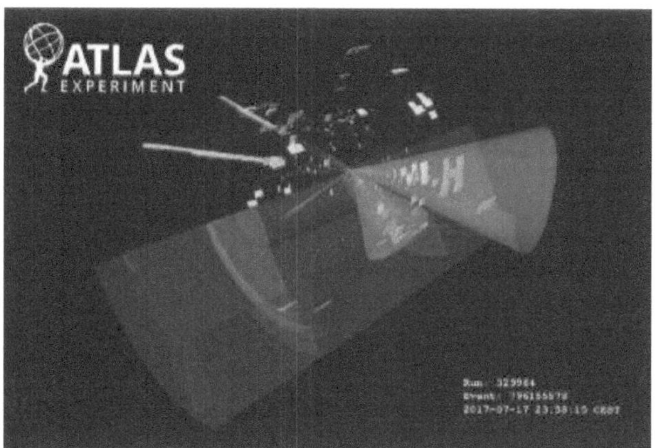

Figure 1.12: ATLAS event display of a pair of Higgs boson candidates decaying to two b-quarks and two photons [30]. Charged particle tracks are shown in green and energy deposits in calorimeter cells in yellow. The two candidate b-quark jets are shown as red cones and the two candidate photons as cyan towers.

outward from the calorimeter surfaces.

Different particles create different sets of signals in the detector. For example an electron leaves a track as well as creating an electromagnetic shower. Photons, which also shower in the electromagnetic calorimeter, can be distinguished by the lack of a matching track. As shown in Figure 1.13, many types of particles can be identified by combining information from each of the subdetectors.

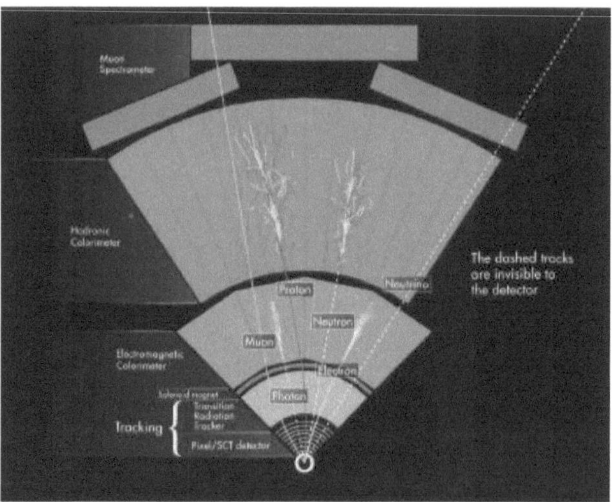

Figure 1.13: Diagram showing the characteristic signature of various SM particles in the ATLAS detector [31].

1.9 Object Definitions

From the tracks and topo-clusters, many physics objects can be defined. Each object represents the full detector response to a single particle or shower of particles. The following list contains only the set of objects used in Chapters 2 and 3 of this book. Each object was itself calibrated to ensure optimal detector and simulation performance.

Muons Muon candidates were reconstructed primarily from tracks in the ID matched to tracks in the MS [32]. Muons are required to pass certain quality criteria based on the number of hits in the detector, the number of 'holes' (i.e. active sensors the track passes through without leaving a hit) and the χ^2 of the track. Several sets of quality criteria are defined for different purposes, with more stringent criteria used to lower rates of false positive identifications, at the cost of larger chances of rejecting real muons. The $g \rightarrow bb$ calibration uses the 'Loose' criteria defined in Ref. [32], while muons for the $HH \rightarrow 4b$ analysis are required to pass the 'Medium' criteria.

Jets As previously mentioned, particle decays involving quarks and gluons result in collimated sprays of hadrons called jets. These jets appear in the detector as nearby or overlapping hadronic showers which can be combined through jet clustering algorithms to reconstruct the kinematic properties of the original source of the jet. The clustering algorithms used by ATLAS belong to the k_T family of sequential recombination algorithms [33, 34]. Motivated by a desire to work backwards through a tree of $1 \rightarrow 2$ particle decays, these algorithms combine pairs of particles sequentially. At each step of the algorithm, a distance metric, d_{ij}, is calculated between each pair of particles and merges the two nearest into a 'pseudo-jet'. A cut-off distance in $\eta - \phi$ space, R, is given as an input parameter to the algorithm and determines the maximum separation at which particles and/or 'pseudo-jets' can be merged. The algorithm runs until no further merges are possible, and the final set of 'pseudo-jets' is returned as the set of jets in the event. The jets used in this book are clustered using the 'anti-k_T' algorithm [35], which has a distance metric,

$$d_{ij} = min(p_{Ti}^{-2}, p_{Tj}^{-2})\frac{(\eta_i - \eta_j)^2 + (\phi_i - \phi_j)^2}{R^2}. \tag{1.12}$$

The algorithm preferentially merges objects with high p_T and the resulting jets are compact and generally circular. The k_T algorithm, by constrast, merges low p_T objects first with a distance metric of

$$d_{ij} = min(p_{Ti}^{2}, p_{Tj}^{2})\frac{(\eta_i - \eta_j)^2 + (\phi_i - \phi_j)^2}{R^2}. \tag{1.13}$$

Three sets of jets were used. "Small-R" jets were constructed from topo-clusters with $R = 0.4$ and used in the trigger for the $g \rightarrow bb$ calibration as described in Sec. 3.3. The jets used were required to have $p_T > 60\,\text{GeV}$, $|\eta| < 2.5$, and pass a jet vertex tagging selection [36] designed to remove jets from pileup vertices. "Large-R" or "fat" jets were also constructed from topo-clusters and were used for objects with 2-pronged decays such as Higgs or vector bosons. Large-R jets used $R = 1.0$, and cover a significant portion of the detector, so they pick up more energy from pileup and the underlying event than small-R jets. An extra trimming step [37] was used to mitigate this: the final jet constituents were reclustered into $R = 0.2$ subjets using the k_T algorithm, and subjets whose p_T was less than 5% of that of the large-R jet were removed [38]. Trimmed large-R jets were required to have $p_T > 250\,$GeV and $|\eta| < 2.0$ to be fully contained in the central region of the detector.

Finally, a set of variable-radius 'track-jets' were used for the b-tagging, as described in Section 1.10. These jets were clustered from tracks with the anti-k_T algorithm and a p_T-dependent R parameter [39],

$$R \equiv \frac{\rho}{p_T}. \tag{1.14}$$

The size of these jets decreases with higher p_T, as more energetic decays are expected to be more collimated. The value of ρ used, $\rho = 30\,$GeV, was optimized to maintain truth-level[3] double-b identification efficiency across the full range of Higgs boson jet p_T [40]. The same optimization was performed to determine the minimum and maximum values of the R parameter, $R_{min} = 0.02$ and $R_{max} = 0.4$.

Track-jets were required to have $p_T > 7\,$GeV and $|\eta| < 2.5$. Track-jets were matched to large-R jets through ghost association [41], which simulates whether an object would be clustered into a jet by adding large numbers of zero-momentum 'ghost' particles to the anti-k_T algorithm. As the 'ghosts' have no effect on objects they merge with, they can be added in arbitrarily large numbers and the area of a jet can be defined as the area containing all ghosts merged into the jet.

1.10 Flavor Tagging

The identification of jets containing bottom quarks, referred to as b-jets, is a topic of special interest for the ATLAS physics program. Many analyses, such as the $HH \rightarrow 4b$ analysis presented in this book, want to separate b-jets from much more common light jets, e.g. jets containing only light quarks. A set of specialized algorithms used to identify b-jets in the ATLAS detector, referred to as b-tagging algorithms or b-taggers, have been under continuous development since ATLAS was built. These algorithms rely on some unique properties of the decays of b-hadrons that ID was

---

[3]Truth-level refers to particle information provided by the Monte Carlo simulations described in Sec. 1.11 before the detector simulation is applied.

designed to exploit.

1.10.1 Properties of b-jets

The b quark is especially interesting both for the study of the b-mesons and because it is preferentially produced in the decays of top quarks and of Higgs bosons. With a mass of 4.18 GeV [3], the b quark is the second-heaviest fermion in the Standard Model. Bottom quarks decay through the weak force, almost exclusively to charm quarks. Due to the rare nature of these processes, the b-hadrons have relatively long lifetimes, $\tau \approx 1.5$ ps, and large decay lengths, $\langle c\tau \rangle \approx 450\,\mu$m [3]. The mean flight length of b-hadrons produced in LHC collisions is significant, with b-hadrons often travelling through the first pixel layer of the ATLAS inner detector before decaying. This lifetime is used to identify b-hadron decays in several ways, as shown in Figure 1.14.

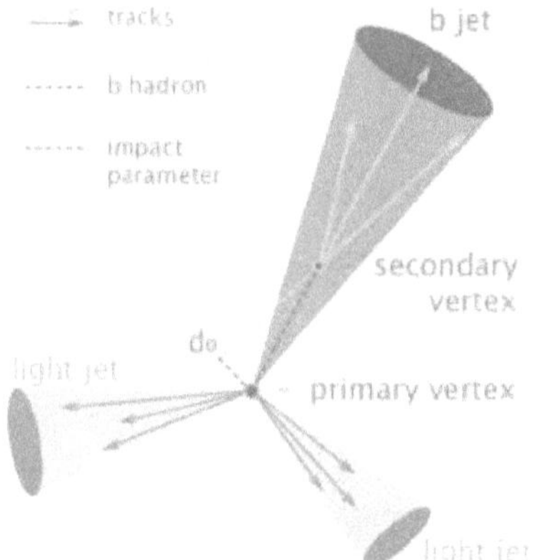

Figure 1.14: Schematic diagram of a jet from the decay of a b-hadron, compared to a jet containing only light hadrons. The secondary vertex is characteristic of b-jets due to the relatively long lifetime of the b-hadron. Tracks coming from such secondary vertices typically have large impact parameters, defined as the distance of closest approach to the primary collision vertex.

The most direct method is to search for signs of a secondary decay vertex, i.e. two or more tracks coming from a single point away from the primary collision vertex. Secondary vertex reconstruction provides a distinctive signature of long-lived particles, but requires accurate reconstruction of all tracks involved. A simpler method is to look for individual tracks passing close to, but not coming from, the collision vertex. The distance of closest approach between a track and the primary vertex is called the 'impact parameter' (IP), and separate measurements are used for the longitudinal (z_0) and transverse (d_0) components. Tracks with large impact parameters can be 'faked' by tracks coming from pile-up collisions or by mis-reconstructed tracks. Finally, approximately 10% of b decays are semileptonic [3], i.e. $b \rightarrow c\ell\nu_\ell$, producing a lepton in the jet. Muons are produced at higher rates from these decays than in light jet processes, making the presence of a muon a useful identification method independent of lifetime measurements. Checking for muons is, however, less powerful than the lifetime-based methods.

The properties of c-jets, jets containing c hadrons, are between those of b-jets and those of jets containing only u, d, and s quarks (collectively called light-jets). With lifetimes around $\tau \approx 0.5\,\mathrm{ps}$ ($\langle c\tau \rangle \approx 150\,\mu\mathrm{m}$) [3], most c-hadrons decay before reaching the ATLAS detector. c-hadrons produced from b decays can travel far enough that their tertiary decay vertex can be distinguished from the secondary vertex of the b decay. c-hadrons produced in the primary collision, however, generally decay before the first pixel layer and reconstructing secondary vertices is often impossible. c decays can still mimic b decays and measuring the c mis-tag rate is an important part of the study of any b-tagging algorithm.

1.10.2 b-tagging algorithms

Many different b-tagging algorithms have been developed within the ATLAS collaboration, optimizing for different kinematic regimes and different jet reconstruction methods. For high-p_T b-jets, many of these algorithms follow a similar prescription. Small-radius jets are used to resolve individual b-hadron decays, and the algorithm runs on a set of tracks associated with the jet. Algorithms are split into two classes. 'Low-level' algorithms focus on a single aspect of the b-hadron decay, e.g. reconstructing secondary vertices, while 'high-level' algorithms combine multiple low-level algorithms for improved overall performance. For example, the high-level DL1r algorithm [42] is a deep neural network (DNN) using as input the low-level RNNIP [43], SV1 [44], and JetFitter [45] algorithms, along with basic kinematic information about the jet [46]. The low-level algorithms each target different signatures. RNNIP uses a recurrent neural network (RNN) to identify b-jets based on track impact parameters. Impact parameter-based taggers using a log-likelihood ratio discriminant, called IP2D and IP3D [47], have also been used in the past. SV1 instead uses the subset of tracks with large impact parameters to form secondary vertices. The output score is based on the likelihood that a true secondary vertex is found, and the compatibility of the vertex with a b-hadron decay. Lastly, the JetFitter algorithm attempts to reconstruct the full b-hadron decay path, including tertiary vertices from c-hadron decays. A neural network uses several of the variables associated with this reconstuction to provide an output score. Each of these low-level algorithms was developed and trained separately, and provide comparable light jet rejection for any given b-tagging efficiency, as shown in Figure 1.15a. The high-level combinations, DL1r and MV2, both provide improved performance over any of the individual low-level algorithms, also shown in Figure 1.15a [46]. The MV2 algorithm in Figure 1.15 differs from DL1 in that it combines the low-level algorithms using a boosted decision tree rather than a neural network. Both low-level and high-level algorithms are under active development, and significant improvements were made in 2019 by training on and tagging variable-radius track-jets (as opposed to R=0.2 anti-k_T track-jets) and including RNNIP instead of the older IP2D and IP3D algorithms [48]. The improvement of the optimized DL1r 2019 algorithm (the 'r' stands for RNNIP) over the 2018 algorithms is illustrated in Figure 1.15b.

The DL1r network produces three classification scores for each input jet, indicating whether it is b-jet-like, c-jet-like, or light-jet-like (i.e. contains no b or c-hadrons). The final b-tagging score is defined by ratios of the individual classifiers:

$$D_{\text{DL1r}} = \ln\left(\frac{p_b}{f_c p_c + (1 - f_c)p_l}\right), \tag{1.15}$$

where p_b, p_c, and p_l represent the b-jet, c-jet and light jet scores respectively, and f_c is the c-jet fraction in the sample. By removing events with low b-tagging scores, analyses can reject large

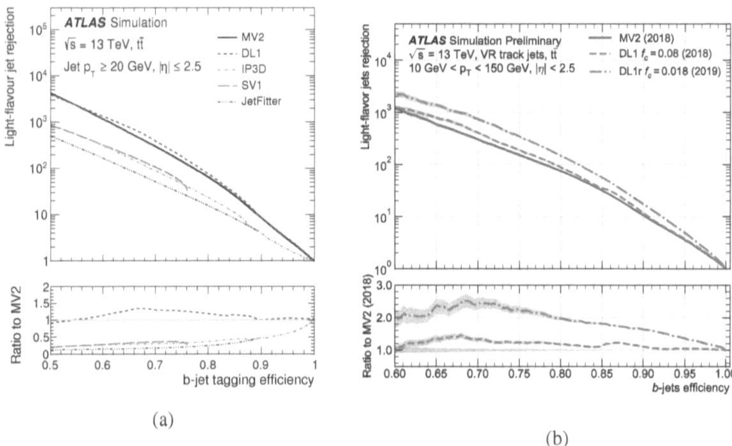

Figure 1.15: Light-jet rejection as a function of b-tagging efficiency in $t\bar{t}$ events. Comparisons are shown for (a) performance of 2018 versions of the low-level IP3D, SV1 and JetFitter algorithms along with the high-level MV2 and DL1 algorithms [46], and (b) the 2018 versions of MV2 and DL1 compared to the 2019 version of DL1r [48].

numbers of light jets while keeping most b-jets for study. Typically, the cut values used are defined by the fraction of b-jets kept in some representative set of simulated events. A cut value with an expected b-tagging efficiency of 70%, for example, would be referred to as the 70% working point (WP). In any analysis, the b-tagging algorithm is applied to both real and simulated data, and a dedicated calibration is needed to ensure the results of these two datasets are compatible. The calibration measures a scale factor (SF) defined as $SF = \epsilon_{\text{data}}/\epsilon_{\text{MC}}$. The scale factor is used to adjust the b-tagging efficiency of the simulation to match that observed in data. The method used to derive scale factors for the DL1r algorithm is described in Ref. [46], while similar calibrations for the mis-tag scale factors for c-jets and light jets, are described in Refs. [49, 50] respectively.

1.11 Datasets Used

All results in this book use the full ATLAS Run 2 dataset, consisting of 139 fb^{-1} of $\sqrt{s} = 13$ TeV proton-proton collision data taken in 2015, 2016, 2017, and 2018. Monte Carlo (MC) simulations were used to augment collision data and test for signals of BSM theories. Simulation was done in a three-step process consisting of event generation, parton showering and hadronization, and detector reconstruction, with separate programs used for each. In all cases, three sets of simulations were produced to replicate the data-taking conditions of 2015+2016, 2017, and 2018 separately.

Feynman diagrams for the $HH \to 4b$ signal models were evaluated at leading-order (LO) in α_S using MADGRAPH [51]. The scalar model was implemented as a two-Higgs-doublet model where the new neutral scalar was produced through gluon-gluon fusion and forced to decay to a pair of Higgs bosons. The scalar width is assigned to be much smaller than the detector resolution, and no other BSM processes were considered in the production. The spin-2 resonance was implemented in the Randall-Sundrum model with the parameter $k/\overline{M}_{Pl} = 1$. In both cases, additional particles predicted by the model do not affect the calculations. Parton showering and hadronization for the spin-0 samples were done using HERWIG 7 [52] with EVTGEN [53] for modelling heavy flavor decays. The MMHT2014 at LO [54] parton distribution funtions (PDFs) was used for the parton shower, and the underlying event was modelled using the default HERWIG 7.1 parameters. The spin-2 samples used PYTHIA 8 [55] with EVTGEN for parton showering and hadronization. The A14 [56] set of ATLAS tuned parameters were used for the underlying event, and the NNPDF2.3 at LO [57] PDF set. Full simulation of the interactions of particles with the ATLAS detector was done with GEANT 4 [58]. Events were generated for resonant masses ranging from 900 GeV to 5 TeV for both signal models. The full list of resonance masses used, as well as the number of events generated for each, is given in Table 1.1.

In addition to the signal hypotheses, three sets of SM processes were simulated. PYTHIA 8 was used to simulate $2 \to 2$ QCD interactions at LO in α_S to model multijet processes. While only $2 \to 2$ matrix elements were evaluated, the parton shower model used by PYTHIA 8, with EVTGEN, includes gluon radiation and splitting that can result in additional jets. The $g \to bb$ calibration uses two sets of multijet MC: one inclusive sample and one sample where the events were required to contain a muon. These samples were generated independently. In addition, $t\bar{t}$ processes were simulated to model their contribution to the $HH \to 4b$ background. $t\bar{t}$ event generation is done at next-to-leading-order (NLO) in α_S using POWHEG-BOX 2 [59, 60, 61, 62], with PYTHIA 8 and EVTGEN for parton showering and hadronization. The POWHEG damping parameter h_{damp}, which affects the modelling of radiation, is set to the value observed to best model the data, 1.5 times the top quark mass [63].

	Events generated			Events generated	
Signal mass [GeV]	G^*_{KK}	X	Signal mass [GeV]	G^*_{KK}	X
900	225k	190k	2000	345k	70k
1000	255k	158k	2250	255k	-
1100	253k	70k	2500	161k	70k
1200	255k	70k	2750	165k	-
1300	75k	70k	3000	305k	69k
1400	254k	70k	3500	119k	-
1500	253k	70k	4000	120k	70k
1600	255k	68k	4500	119k	-
1800	65k	70k	5000	120k	70k

Background sample	Approximate events generated
$t\bar{t}$	530M
Multijet (inclusive)	200M
Multijet (filtered)	51M

Table 1.1: Number of events generated for each simulated sample. Far more simulation is required to accurately model the background processes than for the resonant signals.

Chapter 2: $HH \rightarrow 4b$ Analysis

2.1 Search Overview

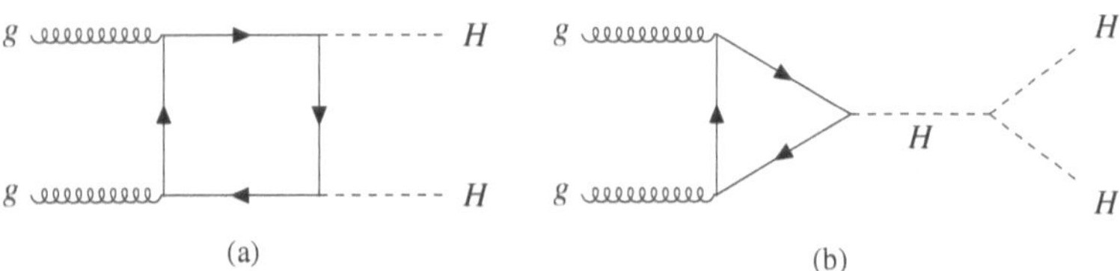

Figure 2.1: Feynman diagrams for Higgs boson pair production via gluon–gluon fusion in the Standard Model. The (a) box and (b) triangle diagrams interfere destructively resulting a SM cross-section of $\sigma_{HH} = 31.05^{+6\%}_{-23\%}$ fb at the LHC [64, 65].

The $HH \rightarrow 4b$ analysis searches for the BSM production of pairs of Higgs bosons, through the $b\bar{b}b\bar{b}$ decay channel. Pairs of Higgs bosons can be produced through gluon–gluon fusion via the processes shown in Figure 2.1, referred to colloquially as the box and triangle diagrams. However, these diagrams interfere destructively, suppressing the SM cross-section to just $\sigma_{HH} = 31.05^{+6\%}_{-23\%}$ fb at the LHC center-of-mass energy of 13 TeV [64, 65] [1]. Many BSM scenarios predict changes to the Higgs sector and specifically heavy resonances that can be seen by the ATLAS detector. Among these are the two benchmark models used for this analysis, the 2HDM and RS models described in Section 1.3. Both predict new heavy resonances with large branching ratios to pairs of Higgs bosons, greatly increasing σ_{HH} around the resonance mass. Both signal hypotheses can also be reinterpreted to calculate discovery potential or limits on similar signals in other BSM models. This analysis aims to discover either a spin-0 scalar or spin-2 graviton or, failing that, to set improved limits on the cross-section of such resonances.

---

[1] The cross-section presented here is calculated using the NLO-improved NNLO approximation. The cross-section has been calculated to NNLO using the Higgs Effective Field Theory approximation, $m_{\text{top}} \rightarrow \infty$, and the full NLO result is corrected by the ratio between HEFT NNLO and NLO predictions.

The ATLAS and CMS [66] experiments at CERN have searched for Higgs boson pair production in a variety of decay channels. The ATLAS collaboration has set limits on resonant HH production in the boosted $b\bar{b}\tau^+\tau^-$ final state [67] using the full 139 fb^{-1} Run 2 dataset, as well in the $b\bar{b}b\bar{b}$ [68, 69], $b\bar{b}\tau^+\tau^-$ [70], $b\bar{b}\gamma\gamma$ [71], $b\bar{b}W^+W^-$ [72], $W^+W^-\gamma\gamma$ [73], and $W^+W^-W^+W^-$ [74] final states using the first 36.1 fb^{-1} of Run 2 data. A combination of the partial Run 2 results was also performed [75]. The CMS collaboration has similarly set limits on the $b\bar{b}b\bar{b}$ [76], $b\bar{b}\gamma\gamma$ [77], $b\bar{b}\tau^+\tau^-$ [78], $b\bar{b}\ell\nu\ell\nu$ [79], and $b\bar{b}ZZ$ [80] final states, and a combination of these [81] using 35.9 fb^{-1} of Run 2 data. No significant deviations from the SM have been observed by either collaboration.

The various HH decay channels have different advantages and disadvantages in the size of the expected signal and the size and complexity of the backgrounds. The motivation for the $b\bar{b}b\bar{b}$ channel comes from its large branching fraction. The Higgs boson decays to $b\bar{b}$ pairs with a branching fraction of around 58%, resulting in a 34% branching fraction of $HH \rightarrow 4b$. The relative branching fractions of the largest final states are shown in Figure 2.2, with the highlighted states showing which are used in ATLAS analyses. The disadvantage of the $b\bar{b}b\bar{b}$ channel is the large multijet background, which must be modelled using data-driven techniques due to the difficulty of the simulating pure QCD processes. b-tagging algorithms are used to reduce the contribution of light jets, though the remainder still forms the bulk of the background along with the irreducible $g \rightarrow bb$ processes. Hadronic decays of pair-produced top quarks also contribute significantly. The search presented in this book uses recent improvements in b-tagging and updated background modelling techniques, as well as the full 139 fb^{-1} Run 2 dataset, to improve on previous results.

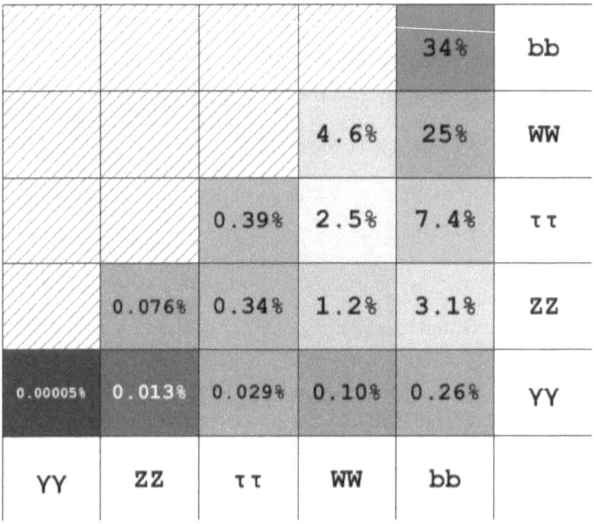

	YY	ZZ	ττ	WW	bb	
					34%	bb
				4.6%	25%	WW
			0.39%	2.5%	7.4%	ττ
		0.076%	0.34%	1.2%	3.1%	ZZ
	0.00005%	0.013%	0.029%	0.10%	0.26%	YY
	YY	ZZ	ττ	WW	bb	

Figure 2.2: HH branching ratios to the final states searched for by the ATLAS and CMS collaborations. The $b\bar{b}b\bar{b}$ channel has the largest branching ratio at 34%.

The $HH \rightarrow 4b$ analysis is split into two kinematic regimes, characterized by the mass of the signal resonance. The methods used for the two regimes are similar but fully independent, with the 'resolved' method used for signal masses of 251-1500 GeV and the 'boosted' method used for mass of 900-5000 GeV. In the overlap region, 900-1500 GeV, both methods are used and the results are combined to increase discovery and limit-setting potential. The focus of this book is on the boosted regime. An overview of the analysis strategy, including brief descriptions of the methods used in both regimes, is presented in Section 2.2. The methods used for the boosted regime are then described in further detail, with the event selection in Section 2.3 and the background modelling in Section 2.4. Treatment of systematics uncertainties are then discussed in Section 2.5. The final fits used to calculate significance and set limits are described in Section 2.6, and the results are shown in Section 2.7. The statistical combination of the resolved and boosted regimes, and the combined results are discussed in Section 2.8.

2.2 Analysis Strategy

The kinematics of $HH \rightarrow 4b$ decays depends on the invariant mass of the Higgs boson pair. In the rest frame of each Higgs boson, it decays into a pair of back-to-back b quarks, which form jets in the manner outlined in Section 1.5. When the rest frame of the Higgs has a small boost relative to the detector frame, these jets are well separated and are reconstructed as such. However, at large boosts they become collimated and are reconstructed as a single jet. The momentum of a Higgs boson produced in resonant decay is about half the difference between the rest mass of the resonance and the rest mass of the Higgs boson pair, 250 GeV. As a rule of thumb, the ΔR distance between the products of two-body decay is $\Delta R \sim 2m/p_\mathrm{T}$, where both p_T and m refer to the parent particle. For the decay products of a Higgs bosons to be contained within a $R = 1.0$ anti-k_t jet, therefore, it must have $p_\mathrm{T} \gtrsim 250$ GeV.

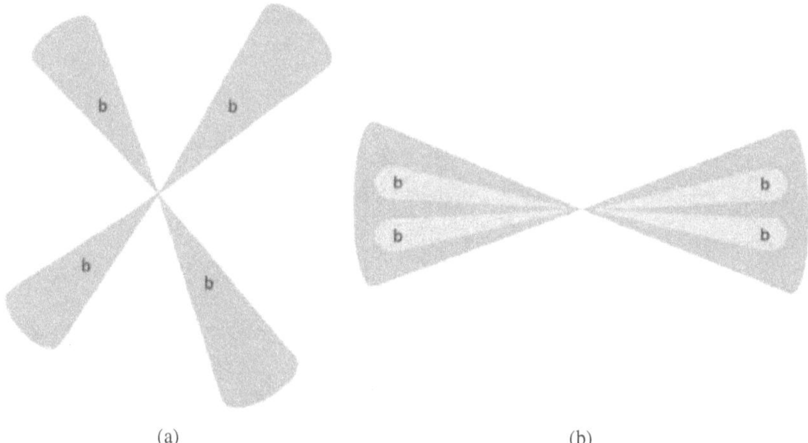

(a)   (b)

Figure 2.3: Topology of (a) resolved and (b) boosted $HH \rightarrow 4b$ decays. Resolved decays are reconstructed as four $R = 0.4$ jets while boosted decays are reconstructed as two $R = 1.0$ jets with associated variable-R track-jets.

In principle, the 'boosted' regime is useful only for finding resonances with mass greater than 750 GeV. In practice, the threshold is above 1000 GeV due to trigger thresholds and reconstruction inefficiencies. On the other hand, the low boost 'resolved' regime can search for resonances with mass just above 250 GeV, but the $R = 0.4$ jets used in this regime lose reconstruction efficiency for bosons with $p_T \gtrsim 1000$ GeV. Again, the thresholds in practice do not match these naive estimates exactly, and the limits of the resolved analysis are competitive with those of the boosted analysis up to 1400 GeV. The resolved analysis is used to search for resonant masses from 251 GeV to 1500 GeV, while the boosted analysis is used to search for resonant masses from 900 GeV to 5000 GeV. In the overlap region, 900-1500 GeV, both analyses capture some events the other would miss and combining the two gives improved sensitivity.

There are many parallels between the analysis strategies used for the resolved and boosted regimes. The high $H \to bb$ branching ratio that motivates this search channel comes at the cost of a large QCD background. In both kinematic regimes the primary challenges of the analysis are to reduce the complex QCD background, and to accurately model the remainder. Background modelling in both cases is done using data-driven techniques, i.e. using data with similar kinematics to the signal region, but with differing b-tagging requirements. Many of the details, however, are different.

The resolved analysis reconstructs Higgs candidates from pairs of $R = 0.4$ jets, using a Boosted Decision Tree (BDT) to determine the optimal pairing for each event. Events are categorized based on the number of b-jets, and the mass of the Higgs candidates. A Neural Network (NN), trained in a dedicated control region, defines the background model by using events with two b-jets to estimate the distributions of those with four b-jets. The primary limitation in the analysis comes from uncertainties on the NN background estimate. Improvements compared to previous ATLAS $HH \to 4b$ analyses come from the BDT pairing algorithm and NN background model, as well as the new 'particle-flow' jet construction technique [82], updated b-tagging algorithms and use of the full Run 2 dataset. The methods used for the boosted search will be presented in further detail in this book. Briefly though, the boosted analysis reconstructs Higgs candidates from individual $R = 1.0$ jets and applies b-tagging to matched variable-radius track-jets. Events are categorized based on the number of b-tags, and the mass of the Higgs candidates. Three separate signal regions are defined based on the number of b-tags, each with a QCD background model derived from a similar region with fewer b-tags. Dedicated control regions are used to improve the background model by measuring the size of the background as well as correcting for biases introduced by differing b-tag requirements. The sub-dominant $t\bar{t}$ background is estimated from MC simulation. The primary limitation of the analysis comes from lack of data at high masses, though uncertainties on the background model are important as well. Improved limits over previous iterations come from updates to the background modelling techniques as well as the increase in data, use of variable-radius track-jets and improved b-tagging algorithms.

2.3 $HH \rightarrow 4b$ Event Selection and Categorization

2.3.1 Object Definitions

To reconstruct $H \rightarrow bb$ decays, anti-k_T $R = 1.0$ jets are built from locally-calibrated topo-clusters and trimmed to remove pile-up contributions. These jets are matched to variable-radius track-jets using ghost-assocation. In addition, there are a two corrections made to improve the jet mass resolution. The first correction accounts for energy lost when muons are produced in b-hadron decays. The production of a muon and muon neutrino reduces the energy deposited in the calorimeter, as the muon deposits relatively little energy and the neutrino is not detected at all. b-jets containing muons, therefore, typically have lower reconstructed momentum than b-jets not containing muons. To correct for this, the energy of the muon track is added to the matched large-R jet. The muons used for the muon-in-jet correction are required to pass the 'medium' identification criteria defined in Ref. [32] and have $p_T > 4\,\text{GeV}$. The correction is applied if a muon is within $\Delta R = \min\left(0.4, 0.04 + 10\,\text{GeV}/p_T^{\text{muon}}\right)$ of one of the b-tagged track-jets associated to the large-R jet. If there are multiple muons within a single track-jet, only the one with the highest p_T is used for the correction.

The second correction uses matched tracks to improve the jet mass calculation. The jet mass is calculated using the combined mass method [83], which makes use of both the calorimeter-based mass calculation, m_{calo}, and the track-assisted mass, m_{TA}, defined as:

$$m_{\text{calo}} = \sqrt{\left(\sum_{i \in \text{Jet}} E_i\right)^2 - \left(\sum_{i \in \text{Jet}} \vec{p}_i\right)^2}$$

$$m_{\text{TA}} = \frac{p_{\text{T}}^{\text{calo}}}{p_{\text{T}}^{\text{track}}} \cdot m_{\text{track}}$$

(2.1)

where E_i and $\vec{p}_i$ are the energy and momentum of the ith topo-cluster constituent of the jet, and m_{track}, $p_{\text{T}}^{\text{track}}$, and $p_{\text{T}}^{\text{calo}}$ are the jet mass and p_{T} calculated from the four-vector sum of all tracks associated to the large-R jet. The muon-in-jet correction is accounted for in the value of m_{calo} used. The combined mass is finally calculated as $m_{\text{comb}} = w \cdot m_{\text{calo}} + (1 - w) \cdot m_{\text{TA}}$, where w is a weight calculated for each large-R jet from the resolution of the calibrated track and calorimeter mass terms. As the track and calorimeter mass terms are only weakly correlated, no correlation terms are required in the linear combination [83]. After calculating the combined mass, the combined jet momentum is recalculated using the calorimeter-based energy measurement, $p_{\text{comb}}^2 = E_{\text{calo}}^2 - m_{\text{comb}}^2$. These corrections reduce the width of the observed Higgs boson mass peak and shift it closer to the known Higgs boson mass of 125 GeV, as shown in Figure 2.4.

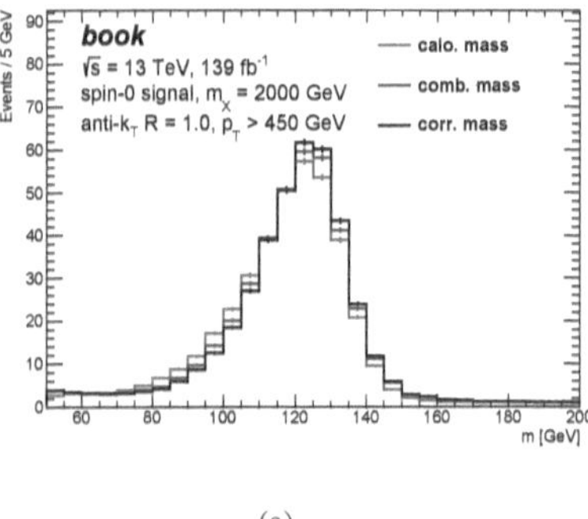

(a)

	Mean [GeV]	Width [GeV]
calo. mass	121.5	35.1
comb. mass	123.6	34.1
corr. mass	124.3	33.8

(b)

Figure 2.4: Large-R jet mass distributions for the 2000 GeV spin-0 signal. Three mass definitions are shown: the calorimeter mass, the combined mass, and the combined mass with a muon-in-jet correction applied. The mean and width of each mass peak is listed in the table.

Year	Online Jet p_T	Online Jet Mass
2015	360 GeV	-
2016	420 GeV	-
2017	420 GeV	40 GeV
2018	420 GeV	35 GeV

Table 2.1: Online large-R jet p_T and mass thresholds by data-taking year. A trimming algorithm was applied to online jets in 2017 and 2018 but not in 2015 or 2016.

2.3.2 Trigger

As described in Section 1.7.5, ATLAS collision data is only recorded when an event passes one of the trigger selections. The triggers used in this analysis are the lowest unprescaled large-R jet triggers for each year of data-taking. That is, data used for the analysis is collected when an event has an 'online jet' satisfying the HLT requirements. The online jet algorithm used by the HLT corresponds approximately to the $R = 1.0$ anti-k_T jets used for later event reconstruction, while the jets used by the L1 trigger are much simpler and only require a certain total energy within a calorimeter region. Each HLT trigger is evaluated on events passing a L1 trigger with a lower threshold, for example the trigger requiring an HLT jet with $p_T > 420$ GeV is fed by a trigger requiring an L1 jet with $p_T > 100$ GeV. The difference in trigger levels means that all events which could pass the HLT trigger would also pass the L1 trigger, i.e. applying both selections is equivalent to applying only the HLT selection. The difference in jet definitions therefore has no impact on the final selection. Similarly, jet p_T thresholds in the analysis are chosen to minimize the impact of differences between the HLT jets and the final 'offline' jets. This is accomplished by requiring that each event contain an offline jet with $p_T > 450$ GeV and mass > 50 GeV. For simplicity, the same cuts are applied to all data and simulated samples, despite year-to-year differences in the triggers used. Table. 2.1 summarizes the online jet requirements for the each of the triggers used, and Figure 2.5 shows the trigger efficiency as a function of offline jet p_T for each year.

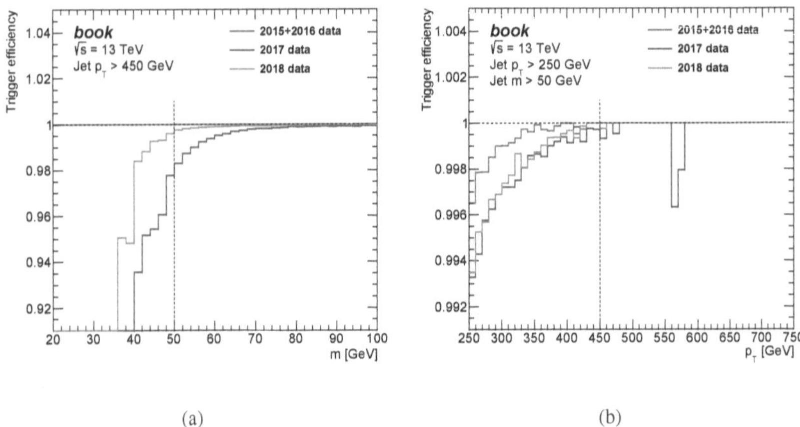

(a)   (b)

Figure 2.5: Efficiency of the lowest unprescaled Large-R jet trigger as a function of (a) jet mass and (b) jet p_T. Efficiency in each variable is measured in a sample with a fully-efficient cut applied to the other variable. The cuts applied in the analysis, mass > 50 GeV and p_T > 450 GeV, respectively, are indicated by vertical lines.

2.3.3 Kinematic cuts

Each event is required to have at least two large-R jets, and these large-R jets themselves are considered to be the "Higgs candidates". The two jets with the highest and second-highest p_T are referred to as the 'leading' and 'sub-leading' Higgs candidates, respectively, while any further large-R jets are ignored for the purposes of this analysis. Both Higgs candidates are required to have p_T > 250 GeV, and as previously mentioned, the leading jet is required to have p_T > 450 GeV to ensure that the triggers are fully efficient. Both Higgs candidates are required to have mass > 50 GeV and be in the central region of the detector, $|\eta|$ < 2.0. The mass distributions in Figure 2.6 show the signals, $t\bar{t}$ background and data with only the trigger cuts applied. Signal jets have reconstructed masses around the 125 GeV Higgs boson mass, while $t\bar{t}$ jets have masses around 80 or 170 GeV, depending on whether the jet contains all the decay products of the top quark or only those of the W boson. Before cuts are applied, the data distribution is swamped by jets from light QCD processes and the mass spectrum shows the tail of a smoothly falling distribution. Figure 2.6

45

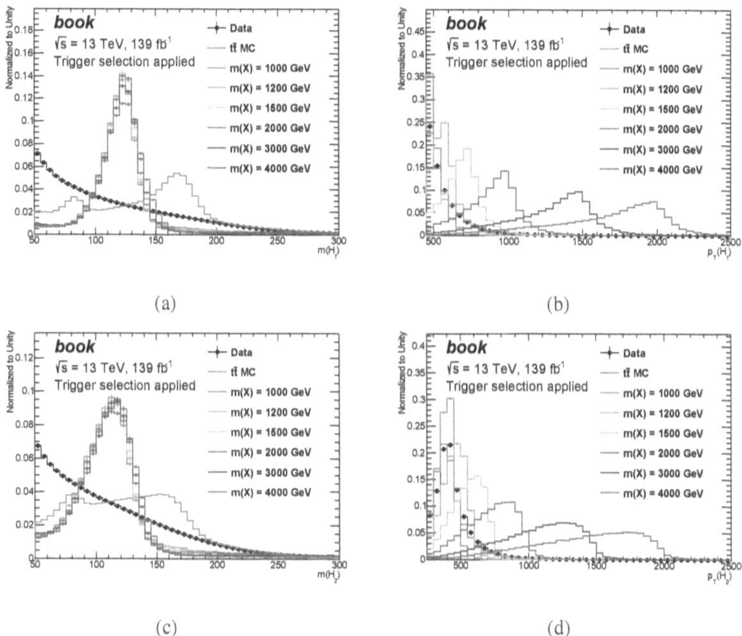

Figure 2.6: Higgs candidate kinematic distributions with only trigger selection cuts applied. The leading Higgs candidate (a) mass and (b) p_T, as well as the sub-leading Higgs candidates (c) mass and (d) p_T for data, $t\bar{t}$, and a range of spin-0 signal masses are shown. The data distribution is dominated by multijet processes, which form the dominant background for the $HH \rightarrow 4b$ analysis.

also shows the p_T distributions at the same stage of the analysis. The p_T distribution of each signal peaks at around half the resonance mass, although the high mass signal distributions have long low-p_T tails.

Finally, a cut on $|\Delta\eta| \equiv |\eta(H_1) - \eta(H_2)| < 1.3$ further ensures the jets are travelling through the center of the detector. Figure 2.7 shows the $|\Delta\eta|$ distributions of the signals, $t\bar{t}$ background and data with only trigger cuts applied. The resonant signals are produced through s-channel processes, whereas the multijet and $t\bar{t}$ components have t- and u-channel components with more forward distributions. While the difference in spin between the two signals leads to a difference in angular distribution, both are more central than the background. The $|\Delta\eta|$ cut is optimized primarily for the scalar signal, but improves the limit for both. After these selections, further event categorization is done based on the mass of the Higgs candidates and the number of b-tagged ghost-associated track-jets.

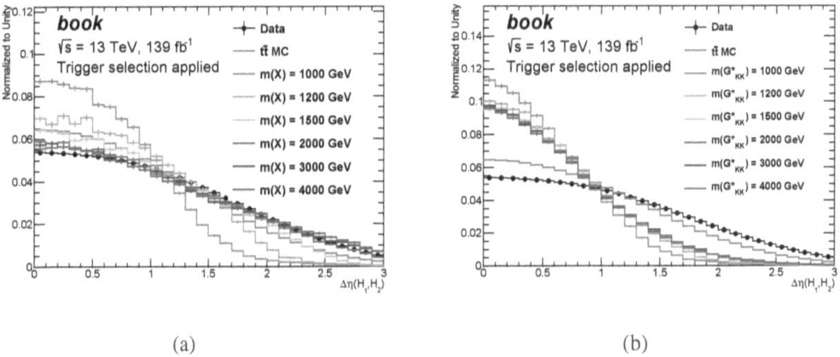

(a)   (b)

Figure 2.7: Distributions of the difference in η between the two Higgs candidates for (a) spin-0 and (b) spin-2 signals, with only trigger selection cuts applied. The data distribution is dominated by multijet processes, which form the dominant background along with $t\bar{t}$ production. The spin-2 signals are more central than spin-0 signals of the same mass.

2.3.4 Mass Plane Regions

Three regions are defined in the 2-D plane of the Higgs candidate masses, called the signal, validation and control regions. Most simulated signal events are contained in the signal region (SR), and this region is used for the limit-setting procedure. The control region (CR) is used to estimate the number of background events in the signal region, through a procedure described in Section 2.4. The validation region (VR), in between the two, is used to calculate an uncertainty on the background estimation method. The following equations are used to define contours in the $m_{H_1} - m_{H_2}$ plane:

$$X_{HH} \equiv \sqrt{\left(\frac{m(H_1) - 124\,\text{GeV}}{0.1 m(H_1)}\right)^2 + \left(\frac{m(H_2) - 115\,\text{GeV}}{0.1 m(H_2)}\right)^2} \tag{2.2a}$$

47

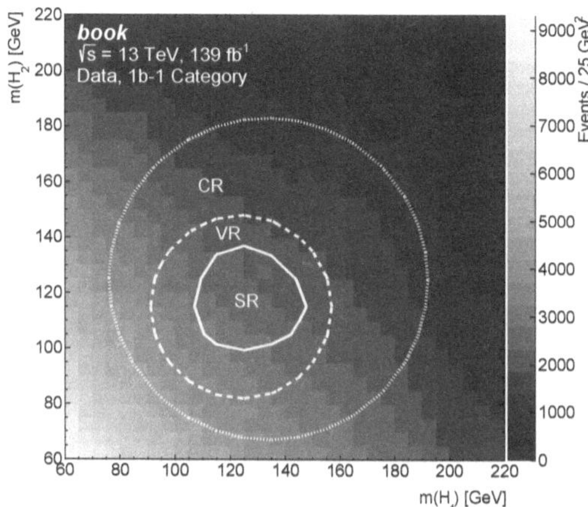

Figure 2.8: Definition of control, validation, and signal regions in the plane of leading and sub-leading Higgs candidate mass (x- and y-axes respectively). The central signal region is surrounded by the validation region, which in turn is surrounded by the control region. The region boundaries are defined by Eq. 2.2. The data shown here comes from the '1b-1' b-tagging channel, as defined in Section 2.3.5.

$$R_{HH}^{VR} \equiv \sqrt{(m(H_1) - 124\,\text{GeV})^2 + (m(H_2) - 115\,\text{GeV})^2} \tag{2.2b}$$

$$R_{HH}^{CR} \equiv \sqrt{(m(H_1) - 134\,\text{GeV})^2 + (m(H_2) - 125\,\text{GeV})^2}. \tag{2.2c}$$

The quantity X_{HH} measures the distance of an event from the Higgs boson mass peak in the $m_{H_1} - m_{H_2}$ plane, and the signal region is defined as the region with $X_{HH} < 1.6$. In addition, a validation region is defined with an outer edge given by $R_{HH}^{VR} < 33\,\text{GeV}$, and a control region is defined by the outer edge $R_{HH}^{CR} < 58\,\text{GeV}$. The inner edge of each region is formed by the region it contains, so the control region ends at the boundary of the validation region and the validation region ends at the signal region. The boundaries of the control, validation, and signal regions are shown in Figure 2.8 in the $m_{H_1} - m_{H_2}$ plane, and summarized in Table 2.2. The control region is shifted to slightly higher masses relative to the validation and signal regions in order to avoid the low mass peak of the QCD distribution. The control region is used for background estimation, as described in Sec. 2.4.2, while the validation region is used to define a systematic uncertainty and to test the limit-setting method. The final search is performed in the signal region.

signal region	validation region	control region
$X_{HH} < 1.6$	$X_{HH} > 1.6$, $R_{HH}^{VR} < 33\,\mathrm{GeV}$	$R_{HH}^{VR} > 33\,\mathrm{GeV}$, $R_{HH}^{CR} < 58\,\mathrm{GeV}$

Table 2.2: Summary of signal, validation and control regions (SR, VR, and CR respectively). The values of X_{HH}, R_{HH}^{VR} and R_{HH}^{CR} are defined in Eq. 2.2.

2.3.5 Tagging Channels

The primary method of separating signal from backgrounds in this analysis is the use of b-tagging. The DL1r b-tagging algorithm is applied to the two highest p_T variable-radius track-jets ghost-associated to each Higgs candidate. The algorithm combines several different methods used to identify long-lived decays of b-hadrons, as decribed in Section 1.10. The algorithm is applied at the 77% efficiency working point, i.e. the selection applied to the b-tagging score has a 77% 'true positive' rate on simulated b-jets.

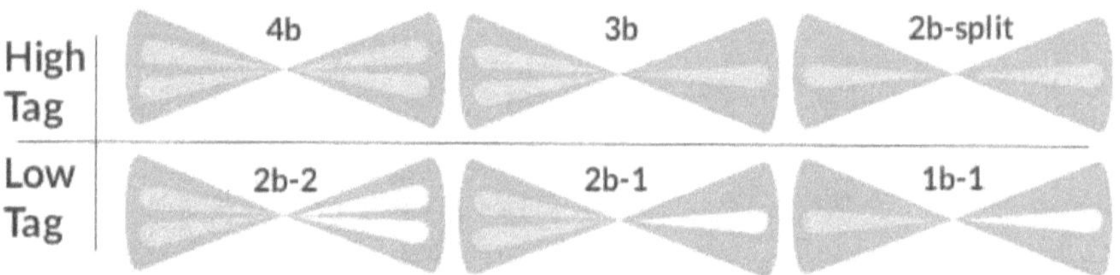

Figure 2.9: Diagram of the three high-tag topologies ($4b$, $3b$ and $2b$-split) with the corresponding channel used to estimate QCD background ($2b$-2, $2b$-1 and $1b$-1) shown directly below. Tagged track-jets are indicated by the small yellow cones within the blue large-R jet cones, while required untagged track-jets are indicated by small white cones.

Events are separated into six independent channels based on the number of b-tags they contain. Events in the '$4b$' channel are required to have four b-tags, two in each Higgs candidate. As $HH \rightarrow 4b$ events contain four b-jets, this channel is the most natural to define and provides the strongest background rejection. Unfortunately, a couple of factors contribute to make the probability of an $HH \rightarrow 4b$ event passing the $4b$ criteria quite small. The first factor is combinatorial: if the probability of a b-jet passing the tagging selection is 77%, then the probability of four b-jets passing is $(0.77)^4 \sim 35\%$. The second is that the efficiency of the algorithm drops for high-p_T Higgs boson jets both because the b-jets become collimated and because the efficiency of the track reconstruction algorithm is reduced. Two additional channels are defined to use events with fewer b-tags: the '$3b$' and '$2b$-split' channels. Events in the $3b$ channel are required to have two b-tags in one Higgs candidate and one b-tag in the other. The $2b$-split channel, meanwhile, is defined by requiring exactly one b-tag in each Higgs candidate. A simultaneous fit is performed in each b-tagging channel on the di-Higgs invariant mass, m_{HH}, and the results are combined to produce limits on both signal models.

For each of these 'high-tag' channels, an additional 'low-tag' channel is defined by reversing the b-tag requirement on one Higgs candidate. The Higgs candidate with no b-tags is required to have a number of track-jets at least equal to the number of b-tags in the corresponding high-tag channel, as illustrated graphically in Figure 2.9. The low-tag partner of the $2b$-split channel, for instance, is labelled '$1b$-1' and requires that one Higgs candidate contain exactly one b-tag while the other contains no b-tags and at least one track-jet. The '$2b$-1' and '$2b$-2' channels, that correspond to $3b$ and $4b$ respectively, each require one Higgs candidate to contain two b-tags and the other to contain none. The difference between $2b$-1 and $2b$-2 is only in the number of track-jets in the untagged Higgs candidate. The $2b$-1 channel requires at least one track-jet in the untagged Higgs candidate while the $2b$-2 channel requires at least two track-jets. The requirements for all channels are summarized in Table 2.3. All events that pass the $2b$-2 criteria necessarily also pass the $2b$-1 criteria. In order to keep these channels statistically independent, events are assigned at random such that 80% are used for the $2b$-1 channel and the remaining 20% are used for the $2b$-2 channel. The increased statistical uncertainties caused by distributing events this way, as opposed to using all events in both channels, has a negligible effect on the final result.

High-tag	4b	3b	2b-split
	2 b-tags	2 b-tags	1 b-tag
	2 b-tags	1 b-tag	1 b-tag
Low-tag	2b-2	2b-1	1b-1
	2 b-tags	2 b-tags	1 b-tag
	> 1 track-jets	> 0 track-jets	> 0 track-jets

Table 2.3: Summary of b-tagging channels. Each channel is defined by two requirements, one on each Higgs candidate. The low-tag channels are listed in the same column as the corresponding high-tag channels.

2.3.6 Resolved Analysis Veto

Separate selections are used for the boosted and resolved $HH \to 4b$ resonant searches. In order to ensure that no events are counted in both, events that pass the resolved signal region selection are removed from consideration for the boosted analysis. Events which pass both selections are events where the Higgs candidates can be reconstructed using pairs of anti-k_T $R = 0.4$ jets and also reconstructed using $R = 1.0$ jets. The resolved veto removes approximately 10% of boosted signal events for resonant masses in the range $m_{HH} = 900\text{-}1300\,\text{GeV}$. However, the impact quickly falls off at higher masses and becomes negligible for $m_{HH} > 1400\,\text{GeV}$. Events removed by this veto are not lost, in the sense that they still contribute to the combined limit through the resolved analysis. The resolved analysis does not apply any vetoes based on the boosted event selection.

2.3.7 Collinear Track-jet Veto

In rare cases, anti-k_T variable radius jets can be constructed such that a high p_T jet is fully contained inside a low p_T jet, shown schematically in Figure 2.10. While the sets of tracks used to construct the jets are well-defined, these cases present problems when training b-tagging algorithms both when determining the set of tracks to use as input to the algorithm, and when applying truth labels for supervised learning. Such events are not used when training the algorithms and, as a precaution, are vetoed from the $HH \to 4b$ search. The veto is only applied if the collinear track-jets in question are both matched to one of the Higgs candidates. While the veto is necessary to ensure a well-defined b-tagging algorithm, it does remove 10% of events for the 1 TeV scalar signal and the effect increases to 20% for the 3 TeV scalars.

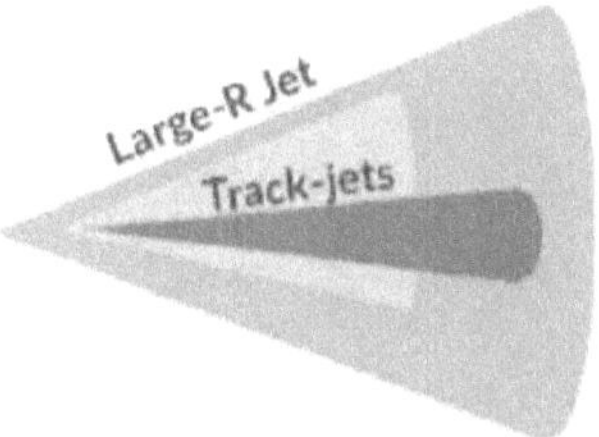

Figure 2.10: Depiction of a large-R jet in which the central axis of a wide, low-p_T track-jet falls within the catchement area of a narrow, high p_T track-jet. Such a configuration causes problems for the b-tagging track-to-jet association algorithm, which assigns tracks to the nearest track-jet in $\eta - \phi$ space.

2.3.8 Cutflow

Table 2.4 shows the effect of each sequential selection cut on the number of expected events for spin-0 and spin-2 signals with a resonance mass of 2 TeV. The production cross-sections are set to 1 fb in both cases. The acceptance $\times$ efficiency, defined as the percentage of total generated events passing the cut, is shown in Figure 2.11. Only around 5-20% of signal events are kept after all cuts are applied, depending on resonance mass. The efficiency of the kinematic cuts, particularly the requirement that the leading Higgs candidate have $p_T > 450$ GeV, have lower efficiency for resonances below 1.2 TeV or so. The efficiency of these cuts plateaus as higher mass resonances produce more boosted jets. At high resonant masses, the efficiency of the b-tagging algorithm is reduced due to merging of the b-jets. As the b-tagging efficiency drops, the number of events in the $2b$-split channel increases relative to $3b$ and $4b$, as shown in Fig. 2.12.

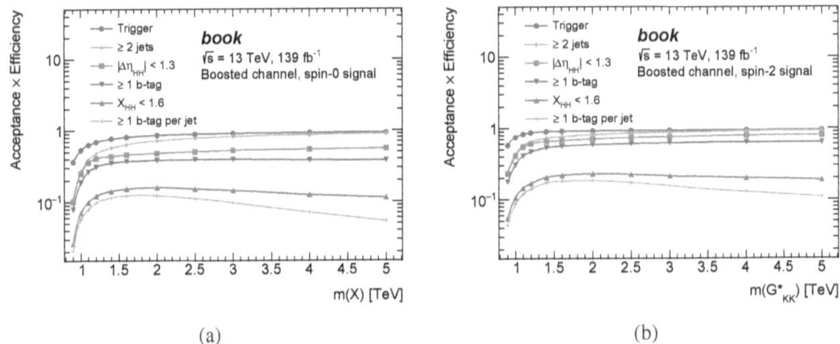

(a)   (b)

Figure 2.11: Acceptance times efficiency as a function of signal resonance mass for (a) spin-0 and (b) spin-2 signals. The values are found by dividing the event yield after each cut by the intial number of events of the respective sample.

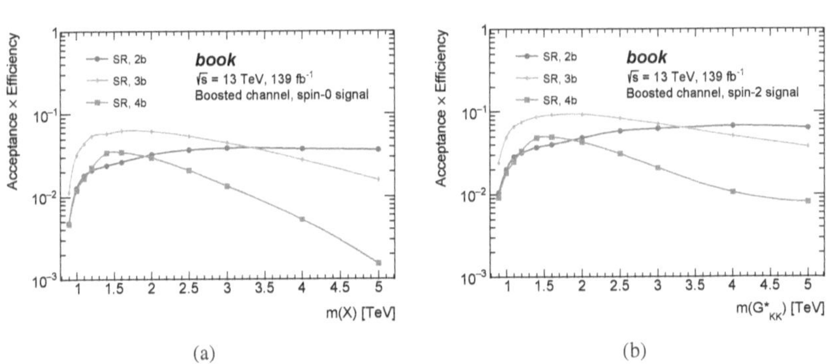

(a)   (b)

Figure 2.12: Acceptance times efficiency as a function of signal resonance mass for (a) spin-0 and (b) spin-2 signals. The values are found by dividing the event yield after the signal region and b-tagging criteria are applied by the initial number of events of the respective sample.

Selection cut		$m(X) = 2\,\mathrm{TeV}$	$m(G^*_{KK}) = 2\,\mathrm{TeV}$
Raw MC events		70000	345000
All events (weighted)		47.1	47.1
Trigger		40.3	43.4
≥ 2 large-R jets		34.7	38.9
Large-R jet mass		34.2	38.4
Leading large-R jet p_T		33.8	38.3
$|\Delta\eta(\mathrm{HH})| < 1.3$		22.7	33.4
Resolved channel veto		22.7	33.3
Collinear track jet veto		18.8	27.6
$4b$	Signal region	1.4	2.3
	Validation region	0.60	1.5
	Control region	0.23	0.83
$3b$	Signal region	2.9	4.4
	Validation region	1.5	2.4
	Control region	0.68	1.1
$2b$-split	Signal region	1.5	2.0
	Validation region	0.90	0.97
	Control region	0.50	0.36

Table 2.4: Efficiency of selection cuts on spin-0 and spin-2 signals with a resonance mass of 2 TeV. Both signals are normalized to a production cross-section of 1 fb.

2.4 Background Estimation

In each b-tagging channel, the primary backgrounds to the $HH \rightarrow 4b$ search are QCD multijet and $t\bar{t}$ events. The relative proportion of these backgrounds depends on the number of b-tags required: in the $4b$ channel, the background is $\sim$ 90% QCD and $\sim$ 10% $t\bar{t}$, but $t\bar{t}$ events make up a larger portion of the $3b$ and $2b$-split backgrounds ($\sim$ 15% and $\sim$ 30% respectively). Other background sources, such as Z+jets and $ZZ \rightarrow b\bar{b}b\bar{b}$, account for <1% of the total and are not considered in the analysis.

A data-driven method is used to estimate the size and shape of the QCD background in each of the $4b$, $3b$ and $2b$-split signal regions. For each b-tagging channel, the shape of the $t\bar{t}$ distributions are taken from Monte Carlo simulation, while the QCD distributions are estimated from data in the corresponding low-tag channels. While the low-tag channels are similar to the high-tag channels, the difference in b-tagging requirements creates some kinematic differences. The iterative reweighting procedure described in Section 2.4.1 is used to correct for these differences in the $1b$-1 and $2b$-1 channels. While a similar kinematic difference is also expected between $4b$ and the corresponding $2b$-2 channel, it is smaller than the statistical uncertainty on the data and therefore ignored. The normalizations of the resulting low-tag distributions are set using maximum likelihood fits in the control region of each b-tagging channel, as described in Section 2.4.2. The di-Higgs invariant mass distributions are then fit to a functional form, as described in Sec. 2.4.3, to produce the final background hypothesis used in the search.

2.4.1 Kinematic Reweighting

To correct for kinematic differences caused by b-tagging the Higgs candidate jets, a reweighting function is applied in the $1b$-1 and $2b$-1 regions. The reweighting procedure uses an iterative spline method, similar to that used in the previous version of this analysis [69]. The method is defined by comparing the kinematics of untagged Higgs candidates in a combined $1b$-1 and $2b$-1 region to the kinematics of tagged Higgs candidates in that same region. The difference between the $1b$-1 and

$2b$-1 regions and their high-tag counterparts, $2b$-split and $3b$ respectively, is only a single tag, i.e. one Higgs candidate has 0 b-tags in the low-tag region whereas one Higgs candidate has exactly 1 b-tag in the corresponding high-tag region. The reweighting procedure, therefore, only needs to act on untagged jets such that they match single-tag jets. With this in mind, only single-tag Higgs candidates are used for the reweighting target distributions. Similarly, only one of the track-jets, chosen at random, is reweighted per untagged Higgs candidate. In single-tag Higgs candidates, the b-jet has approximately equal chance of being the first or second track-jet. Reweighting a random track-jet in each untagged Higgs candidate approximates the same tagging distribution. A set of target distributions are defined based on the kinematics of single-tagged Higgs candidates:

1. p_T of the tagged Higgs candidate,

2. p_T of the b-tagged track jet,

3. η of the b-tagged track jet,

4. ΔR between the leading and subleading track jets (where applicable).

Separate distributions are constructed for the p_T of leading and subleading Higgs candidates, as well as for the p_T of leading and subleading track jets. Equivalent distributions are then defined for the untagged Higgs candidates. If the untagged Higgs candidate has more than one track jet, one of the leading two is randomly chosen as input to the reweighting.

At each iteration of the reweighting, the ratios of tagged to untagged distributions are fit to cubic splines. The weights are then updated according to

$$W_i = W_{i-1} \times [(\Pi_j f_{ij}(x_j) - 1) \times L_i + 1], \tag{2.3}$$

where the functions f_{ij} are the splines evaluated on the kinematic variables x_j at iteration i, and the "learning rate", L_i, controls how much the weight can change with each iteration. With a learning rate of $L_i = 1 - 0.5^i$ the splines converge quickly, within three to four iterations. To ensure good agreement between tagged and untagged distributions a total of ten iterations, shown in Figure 2.13, are used. The learning rate is low for the early iterations to reduce numerical instabilities associated with updating based on multiple highly correlated variables simultaneously. After ten iterations, the final weight for each event is

$$W_f = W_0 \times \Pi_{i=1}^{10} [(\Pi_j f_{ij}(x_j) - 1) \times L_i + 1], \qquad (2.4)$$

where W_0 denotes the initial event weight (which is 1 for data). The reweighting function is derived using only the data sample, but is applied to the $t\bar{t}$ MC as well. This ensures that the low-tag $t\bar{t}$ matches the $t\bar{t}$ component of the low-tag data. Note that since the reweighting applies only in the 1b-1 and 2b-1 regions, the high-tag $t\bar{t}$ background is unaffected. Figure 2.14 shows comparisons of the HH invariant mass distributions before and after reweighting in the 2b-split and 3b control regions. While a similar mis-modelling may be expected between the 4b and 2b-2 regions, it is smaller than the statistical uncertainty in the distribution so no reweighting is applied.

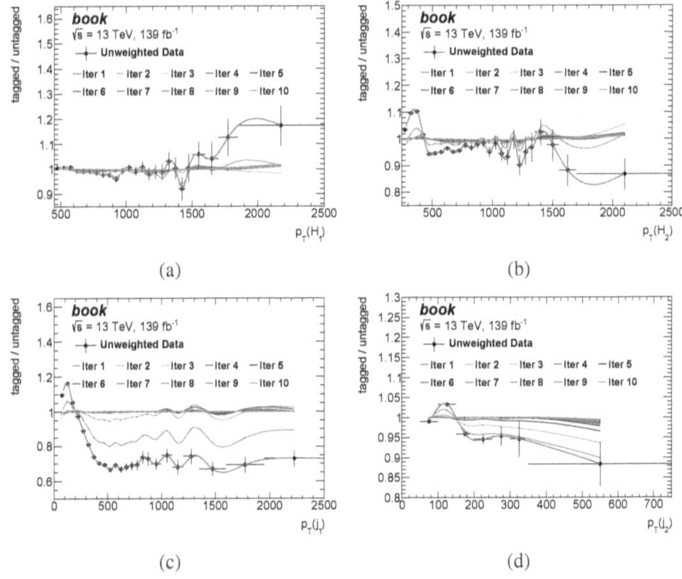

(a) (b)

(c) (d)

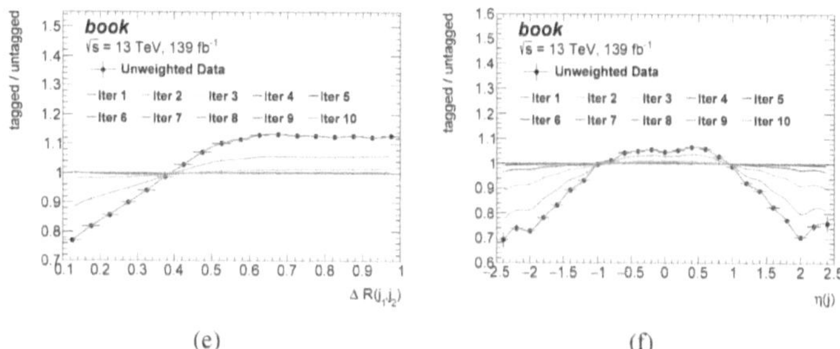

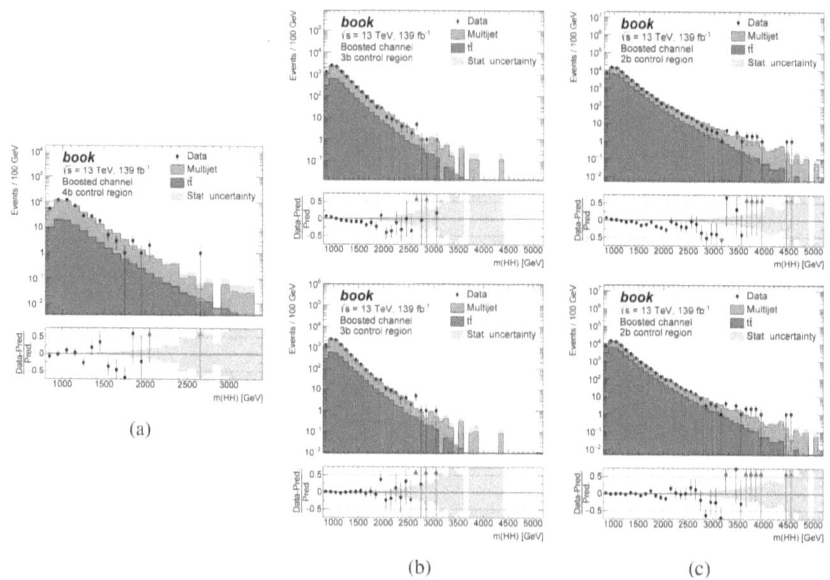

Figure 2.13: Convergence of the spline functions used to reweight the 1*b*-1 and 2*b*-1 regions to correct for differences from the corresponding high-tag channels caused by the *b*-tagging algorithm.

Figure 2.14: *HH* invariant mass in the (a) 4*b*, (b) 3*b* and (c) 2*b*-split control regions. For the 3*b* and 2*b*-split channels, the top plot shows un-reweighted distribution while the bottom plot shows the reweighted distribution. Multijet and $t\bar{t}$ backgrounds are normalized using the μ_{QCD} and $\alpha_{t\bar{t}}$ values defined in Sec. 2.4.2. The gray band shows only statistical uncertainties.

2.4.2 Derivation of Background Normalization

The control regions are used to define the normalization of the background components. In the low-tag channels a model of the QCD background is constructed by subtracting the $t\bar{t}$ MC from the data. This QCD model, along with the high-tag $t\bar{t}$ MC, is then fit to the high-tag data in the control region to determine the normalization factors, μ_{QCD} and $\alpha_{t\bar{t}}$. A two-parameter binned maximum-likelihood fit is used to calculate the scale factors. The high-tag data is fit to the sum of the QCD model and $t\bar{t}$,

$$y^{\text{data},n_b} = \mu_{QCD}\, y^{\text{QCD},n_b} + \alpha_{t\bar{t}}\, y^{t\bar{t},n_b}, \tag{2.5}$$

where n_b indicates the high-tag channel while $n_b - 1$ indicates the corresponding low-tag model and $y^{\text{QCD},n_b} \equiv y^{\text{data},n_b-1} - y^{t\bar{t},n_b-1}$. Assuming uncorrelated Poisson distributions for the data in each histogram bin, and summing over the bins one can derive the likelihood function:

$$\mathcal{L}(\mu_{QCD}, \alpha_{t\bar{t}}) = \prod_{i=1}^{N} e^{-(\mu_{QCD} y_i^{\text{QCD},n_b} + \alpha_{t\bar{t}} y_i^{t\bar{t},n_b})} \frac{(\mu_{QCD} y_i^{\text{QCD},n_b} + \alpha_{t\bar{t}} y_i^{t\bar{t},n_b})^{y_i^{\text{data},n_b}}}{y_i^{\text{data},n_b}!}, \tag{2.6}$$

where the index i runs over bins of the leading Higgs candidate mass, y_i indicating the number of events in bin i for a particular histogram. The fit is performed separately in each b-tagging channel. The final μ_{QCD} is an estimate of the ratio of the number of QCD events in the high-tag channel to the number in the low-tag channel, while the $t\bar{t}$ normalization parameter $\alpha_{t\bar{t}}$, applied after the $t\bar{t}$ is scaled to the total integrated luminosity, is a correction to the MC prediction in this phase space.

The distributions used to calculate the μ_{QCD} and $\alpha_{t\bar{t}}$ normalization factors are shown in Figure 2.15. These distributions are reweighted as described in Section 2.4.1 and the background model is able to match the data even though the $m(H_1)$ distribution is heavily sculpted by the control region contours. In the $4b$ region, $\alpha_{t\bar{t}}$ was found to be essentially unconstrained, due to lack of data, and so it was fixed to one. The fitted values of μ_{QCD} and $\alpha_{t\bar{t}}$ are given in Table 2.5. While the uncertainties on the individual parameters are shown, they can be misleading due to the large correlation between parameters. The correlation between parameters is taken into account in the

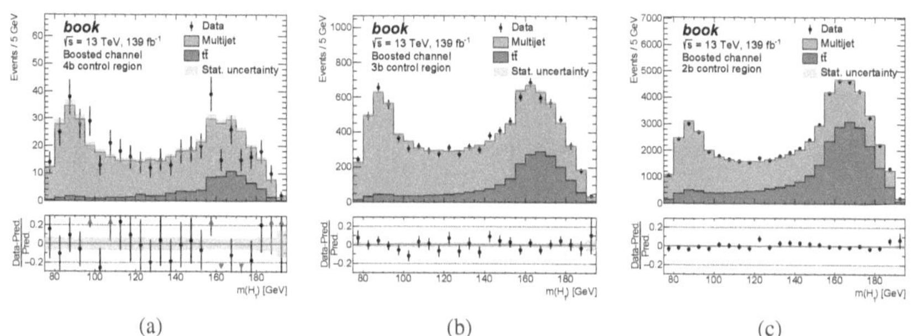

(a)   (b)   (c)

Figure 2.15: Leading Higgs candidate mass distributions in the (a) $4b$, (b) $3b$ and (c) $2b$-split control regions. The control region data is used to determine the normalization of the multijet and $t\bar{t}$ backgrounds. The statistical uncertainty in the fit is shown in the gray bands.

region	$2b$-split	$3b$	$4b$
μ_{QCD}	0.05428 ± 0.00057	0.1201 ± 0.0024	0.0269 ± 0.0015
$\alpha_{t\bar{t}}$	0.827 ± 0.011	0.771 ± 0.041	1
correlation	-0.74	-0.74	-

Table 2.5: Fitted values for μ_{QCD} and $\alpha_{t\bar{t}}$, with statistical uncertainties on the parameters and normalization uncertainties on the backgrounds. μ_{QCD} and $\alpha_{t\bar{t}}$ are used to set the normalization of the background components. In the $4b$ region, $\alpha_{t\bar{t}}$ is fixed to one.

statistical uncertainty of the background model.

An underlying assumption of this method is that these scale factors are roughly constant over the $m_{H_1} - m_{H_2}$ plane, which is verified by using an independent estimate the number of events in the signal region of each b-tagging channel. An extrapolation uncertainty is defined from this estimate, as described in Section 2.5. The fit makes the additional assumption that the $t\bar{t}$ yield given by MC in the low-tag region is correct. The fit is, however, insensitive to small variations in the low-tag $t\bar{t}$ yield so this assumption has no impact on the result.

2.4.3 m_{HH} smoothing fit

In order to reduce the effect of statistical fluctuations at high m_{HH} in the analysis, the multijet and $t\bar{t}$ distributions are fit to the following function:

$$f(x) = \frac{e^{-p_0}}{x^2}(1 - x)^{p_1 - p_2 \ln x}, \tag{2.7}$$

where $x \equiv m_{HH}/\sqrt{s}$ and p_i are dimensionless free parameters. This functional form was chosen from among the so-called "dijet" functions, that have been used to fit falling dijet spectra in similar analyses including the previous round of this analysis [69]. The chosen function was found to be the median in most signal regions among the eight functions tested. A couple of changes were made to improve convergence of the fit: first, the exponential of the p_0 parameter is used rather than p_0 itself, ensuring that the parameters have similar magnitudes. The second changes was to normalize the input distribution to 1 across the fit range during the fitting procedure. After running the fit, the resulting function and associated errors are scaled to the expected number of events.

The fit is performed only in the bins above 1200 GeV to avoid biases from inefficiencies of the boosted selection at lower masses. While the fit range also has an upper limit, set to avoid issues with empty bins, the fitted function is used to smooth over the entire range above 1200 GeV. Due to low statistics in the $4b$ region, the shape of the $t\bar{t}$ distribution in this region is taken from the $3b$ region and scaled to the $4b$ yield. The fit ranges in each of the signal regions are as follows:

- $2b$-split QCD model: 1200-4300 GeV, $t\bar{t}$: 1200-3900 GeV

- $3b$ QCD model: 1200-2800 GeV, $t\bar{t}$: 1200-2200 GeV

- $4b$ QCD model: 1200-2500 GeV

Figure 2.16 shows the fits to the QCD model in each of signal regions while Figure 2.17 shows the fits to the $2b$-split and $3b$ $t\bar{t}$ MC.

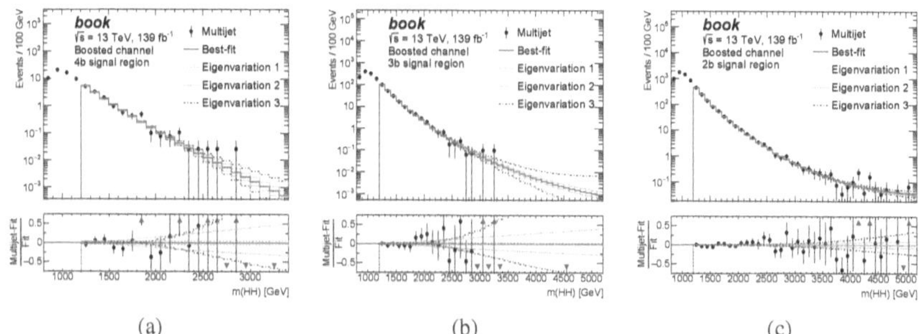

Figure 2.16: Smoothing fit applied to the m_{HH} spectrum of the multijet background model in the (a) $4b$, (b) $3b$ and (c) $2b$-split signal regions. The red curve shows the nominal background estimate, while the blue bands show 'eigenvariations' used to assess the impact of limited statistics on the fit. The eigenvariation method is explained in Section 2.5.

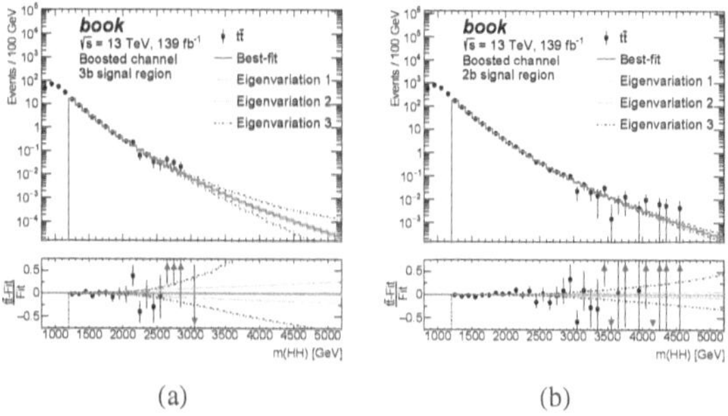

Figure 2.17: Smoothing fit applied to the m_{HH} spectrum of the $t\bar{t}$ background in the (a) $3b$ and (b) $2b$-split signal regions. The red curve shows the nominal background estimate, while the blue bands show 'eigenvariations' used to assess the impact of limited statistics on the fit. The eigenvariation method is explained in Section 2.5.

2.4.4 Background predictions

While defining the background model, data in the signal region was blinded. In order to gain confidence in the model, comparisons between data and prediction in control and validation regions were made using a number of different kinematic variables. The control region plots for two of these variables, the p_T of both Higgs candidates are shown Figure 2.18. These plots include the full background estimation procedure and associated uncertainties. After the kinematic reweighting is applied, good agreement with data is observed for most distributions in all regions.

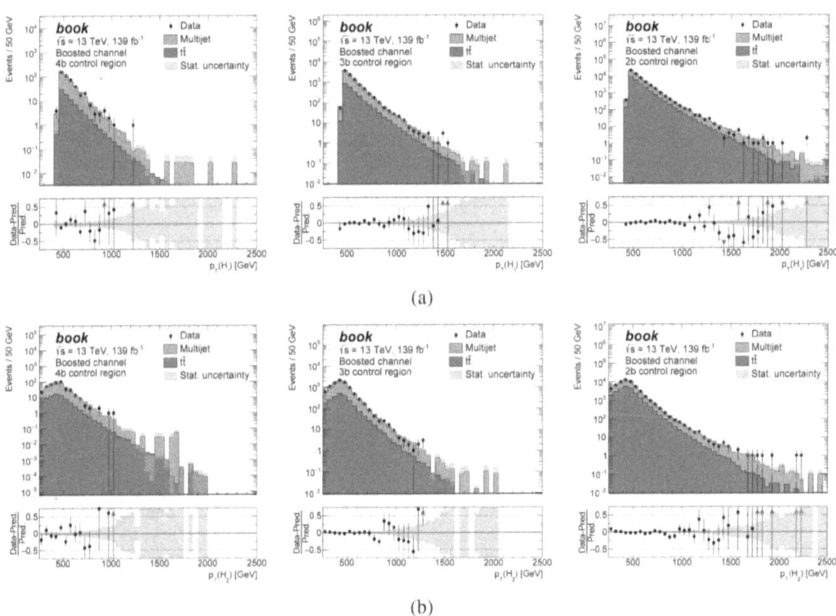

Figure 2.18: The p_T distributions of (a) the leading Higgs candidate and (b) the subleading Higgs candidate in the (left) 4b, (middle) 3b, and (right) 2b-split control regions. The gray band shows the sum of the background modelling uncertainties.

2.5 Systematic Uncertainties

There are a number of sources of systematic uncertainty affecting the boosted $HH \rightarrow 4b$ analysis, the most important of which are uncertainties on the background estimate. Separate systematic variations are derived for uncertainties arising from the methods used to derive the background normalizations and shapes. In addition, two 'non-closure' uncertainties are defined by the discrepancies observed when applying the complete method on an alternate dataset, once using the high-tag validation regions and once using the signal regions in simulated multijet data. These uncertainties are intended to cover any potential biases not explicitly accounted for elsewhere.

2.5.1 Uncertainties on the Background Normalization

Three source of uncertainty are considered for the background normalization: the statistical uncertainty of the fit procedure, the uncertainty associated with the extrapolations from low-tag to high-tag and from control to signal region, and potential for biases due to the definitions of the control regions.

Statistical Uncertainty

The covariance of the fit parameters is found by taking the inverse of the Hessian matrix, $\mathbf{H}$. The Hessian matrix is defined as the matrix of second-derivatives of the likelihood function, i.e. its elements are given by

$$H_{ij} = \frac{\partial^2 \mathcal{L}}{\partial x_i \partial x_j}$$
$$H_{ii} = \frac{\partial^2 \mathcal{L}}{\partial x_i^2}, \tag{2.8}$$

where $\mathcal{L}(\vec{x})$ is the likelihood function and the indices i, j run over the parameters $\vec{x}$. Assuming $\mathcal{L}(\vec{x})$ is approximately Gaussian in the region around the maximum, the inverse Hessian provides a good approximation of the full covariance of the parameter estimates.

The background normalization fit has two free parameters, μ_{QCD} and $\alpha_{t\bar{t}}$, and a 2×2 covariance matrix. The eigenvectors of the covariance matrix define an orthogonal basis of the parameter space, while the eigenvalues define the variance, σ^2, along each axis. Two 'eigenvariations' are created by adjusting the parameters by one standard deviation, σ, along each eigenvector. These eigenvariations change the m_{HH} spectrum of the background hypothesis, by altering the relative proportions of QCD and $t\bar{t}$. The alternate spectra are propogated through the rest of the analysis to provide uncertainty bands on the background estimate. In the $4b$ channel, the fit has only one free parameter and the covariance matrix becomes trivial. Only one eigenvariation is used in this channel.

Extrapolation uncertainty

The normalizations of the QCD and $t\bar{t}$ backgrounds are derived in a control region, and then applied in the signal region. This method relies on the assumptions that the QCD model from the low-tag channels has the same shape as the true QCD in the high-tag channels, and that the scale factors are the same between the control and signal regions. The extrapolation uncertainty provides an estimate of the uncertainties in the background normalization arising from these assumptions.

To assess the extrapolation uncertainty, a Gaussian Process technique is used to interpolate the data in the signal region. The interpolation procedure is done in two steps. First, a fit is performed on the blinded distribution to determine the parameters for a Gaussian two-point correlation function, or kernel. This kernel is then used to predict the values of all the points in the $m_{H_1} - m_{H_2}$ plane, including points in the blinded signal region.

The fit is performed on a data distribution with the $t\bar{t}$ contribution removed, which is expected to have no small-scale structure. The fitted kernel functions are found to have correlation lengths on the order of 100-200 GeV, much larger than the size of the signal region. In addition, the predicted values closely match the actual values outside the signal region. Therefore, the hole in the distribution does not seem to bias the fit, but is properly smoothed over as shown in Figure 2.19. The sparsity of data in the $m_{H_1} - m_{H_2}$ plane in the $4b$ channel requires wider bins to be used in the fit. Still, the extrapolation uncertainty is largest in this channel in part due to low statistics. Note that the QCD distribution used for this procedure can be defined separately for the high and low-tag channels, unlike the QCD model defined by the nominal background estimation technique, which uses low-tag data to define the distribution in the high-tag channels.

A quantitative measure of the extrapolation uncertainty is obtained by calculating the following double ratio:

$$R_{extr} \equiv \frac{N_{SR}^{n_b}/N_{SR}^{n_b-1}}{N_{CR}^{n_b}/N_{CR}^{n_b-1}} \tag{2.9}$$

where N^{n_b} is the number of events in the high-tag channel and N^{n_b-1} is the number of events in the corresponding low-tag channel. The uncertainty, $|R_{extr} - 1|$, is 1.72%, 6.02%, 10.94% in the 2b-split, 3b, and 4b channels respectively.

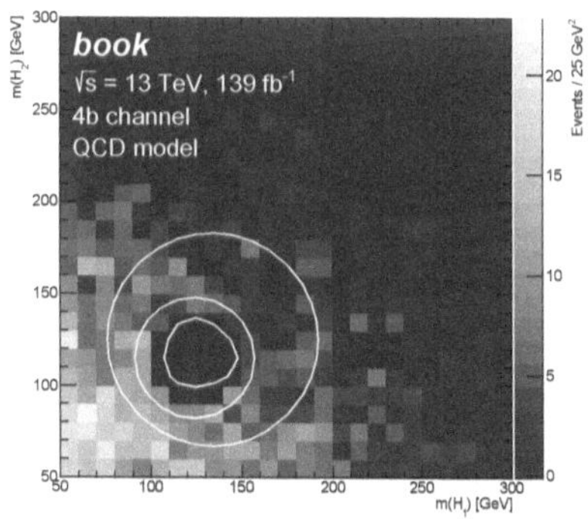

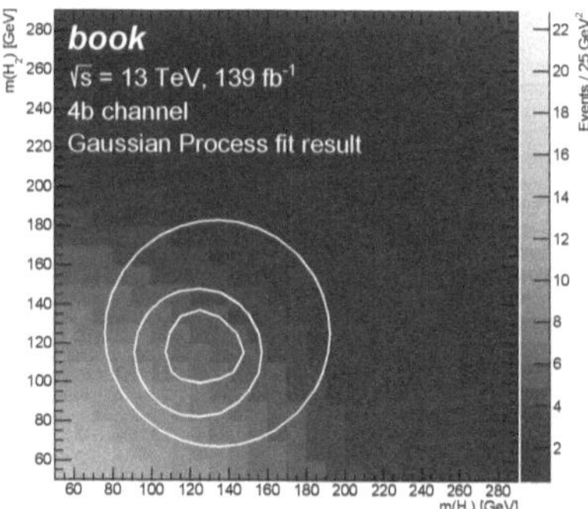

(a)

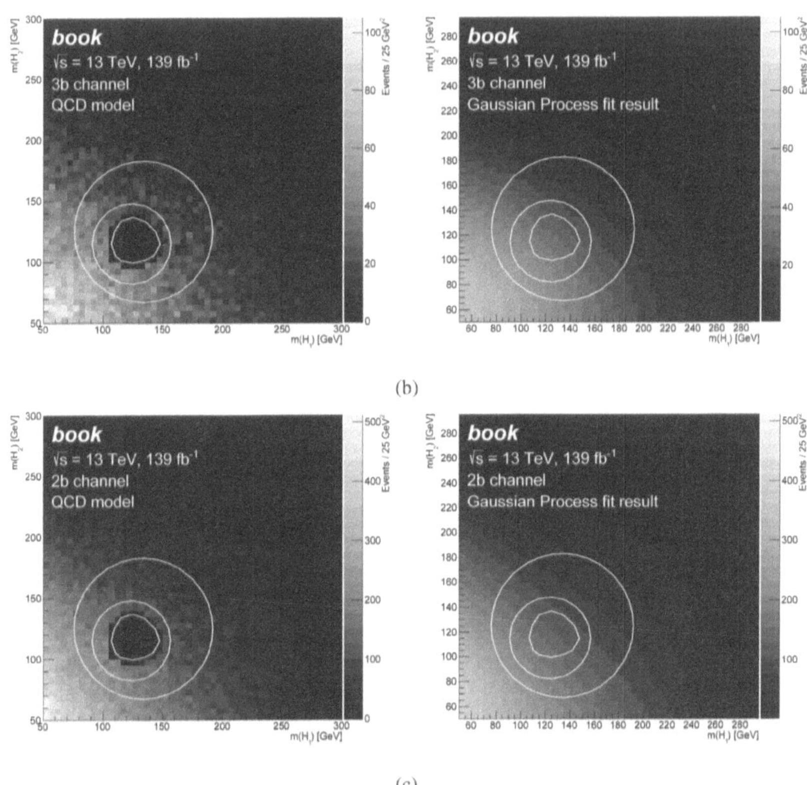

(b)

(c)

Figure 2.19: The result of the Gaussian Process interpolation in the (a) 4b, (b) 3b and (c) 2b-split channels. The multijet model used, data - $\alpha_{t\bar{t}}$ $t\bar{t}$, is shown on the left, while the result of the interpolation is shown on the right. The Gaussian Process is used to assess the uncertainty from the extrapolations from control to signal regions and low-tag to high-tag channels in the background estimate.

Mass-Region Definition Uncertainty

The position and size of the control region used to estimate the background affect the final result, so an uncertainty is added to cover the impact of these choices. The uncertainty is assessed by creating a set of alternate control regions, each defined by some variation of the R_{HH}^{VR} and R_{HH}^{CR} contours of the equations defined in Eq. 2.2. While new validation regions could be defined using the R_{HH}^{VR} variations, only the control region variations are used for the mass-region uncertainty. The background fit is redone for each variation and the background yield is calculated in the signal region. The largest difference in yield between the variations and nominal background estimate is applied as a systematic uncertainty.

Six variations are made, four from moving the centers of the ellipses and two from changing the cut values. The variations are the following:

- **Up-up control region** 3 GeV is subtracted from both the leading and subleading Higgs candidate masses[2]. This moves the center of both the R_{HH}^{VR} and R_{HH}^{CR} circles are moved up and to the right by 3 GeV in the $m_{H_1} - m_{H_2}$ plane.

- **Up-down control region** 3 GeV is subtracted from the leading Higgs candidate mass and added to the subleading Higgs candidate mass. This moves the center of both the R_{HH}^{VR} and R_{HH}^{CR} circles are moved up and to the left in the $m_{H_1} - m_{H_2}$ plane.

- **Down-up control region** 3 GeV is added to the leading Higgs candidate mass and subtracted from the subleading Higgs candidate mass. This moves the center of both the R_{HH}^{VR} and R_{HH}^{CR} circles are moved down and to the right in the $m_{H_1} - m_{H_2}$ plane.

- **Down-down control region** 3 GeV is added both the leading and subleading Higgs candidate masses. This moves the center of both the R_{HH}^{VR} and R_{HH}^{CR} circles are moved down and to the left in the $m_{H_1} - m_{H_2}$ plane.

---

[2]The value of 3 GeV was chosen to be large enough to change the top fraction of the background but not move the control region into the low mass QCD peak, nevertheless it is somewhat arbitrary. The same value was used in previous iterations of the analysis.

- **Large control region** The R_{HH}^{VR} cut is decreased by 3 GeV and the R_{HH}^{CR} cut is increased by 3 GeV. This shrinks the inner boundary while expanding the outer boundary, resulting in a larger control region.

- **Small control region** The R_{HH}^{VR} cut is increased by 3 GeV and the R_{HH}^{CR} cut is decreased by 3 GeV. This expands the inner boundary while shrinking the outer boundary, resulting in a smaller control region.

These variations are done separately for each of the $2b$-split, $3b$, and $4b$ regions and the final region-definition uncertainties are 0.88%, 1.25%, 6.05% respectively.

2.5.2 Uncertainties on the Background Shape

Three uncertainties on the shape of the m_{HH} distribution are defined based on the smoothing method defined in Section 2.4.3. In addition to the uncertainty due to limited statistics, biases can be introduced from the choices of functional form and the range used for the smoothing procedure, so two sets of variations are defined.

Statistial Uncertainty

The statistical uncertainty in the fit itself is accounted for using the same 'eigenvariation' method as is used for the normalization fit. As the smoothing function has three parameters, three variations are created for each fit totaling six for each of the b-tagging channels. The smoothing is done independently on the QCD and $t\bar{t}$ distributions so each variation is uncorrelated to any others.

Choice of Function and Fit Range

In addition to the statistical uncertainty, two sources of systematic uncertainty are identified related to choices made in the fit: one for the choice of functional form and the other for the choice of fit range. The impact of these choices is assessed by making a set of different choices and assuming this set characterizes the space of possible results. The set of alternate functional forms

is given in Table 2.6 and were chosen based on similar sets used in previous publications, e.g.
Refs. [69, 84]. Each functional form is fit to the background model, and the two that differ most
from the nominal 'MJ8' fit are selected to define by an uncertainty band. To be considered for the
uncertainty calculation, a fit must converge and produce a monotonically decreasing curve across
the full m_{HH} spectrum. Similarly, a set of alternate choices of fit range are created by moving
the upper and lower fit bounds independently by 100 GeV (one bin). Of the four variations, the
two that differ the most from the nominal predication are used to define the uncertainty band. The
results of these alternate fits are shown in Figure 2.20 for the QCD model, and Figure 2.21 for $t\bar{t}$.

Name	Functional Form
MJ1	$f_1(x) = e^{-p_0}(1 - x)^{p_1} x^{p_2}$
MJ2	$f_2(x) = e^{-p_0}(1 - x)^{p_1} e^{p_2 x^2}$
MJ3	$f_3(x) = e^{-p_0}(1 - x)^{p_1} x^{p_2 x}$
MJ4	$f_4(x) = e^{-p_0}(1 - x)^{p_1} e^{p_2 \ln x}$
MJ5	$f_5(x) = e^{-p_0}(1 - x)^{p_1}(1 + x)^{p_2 x}$
MJ6	$f_6(x) = e^{-p_0}(1 - x)^{p_1}(1 + x)^{p_2 \ln x}$
MJ7	$f_7(x) = \frac{e^{-p_0}}{x}(1 - x)^{p_1 - p_2 \ln x}$
MJ8	$f_8(x) = \frac{e^{-p_0}}{x^2}(1 - x)^{p_1 - p_2 \ln x}$

Table 2.6: The dijet functions used to fit the background m_{hh} distribution. The functional form
MJ8 was chosen to provide the nominal background estimate.

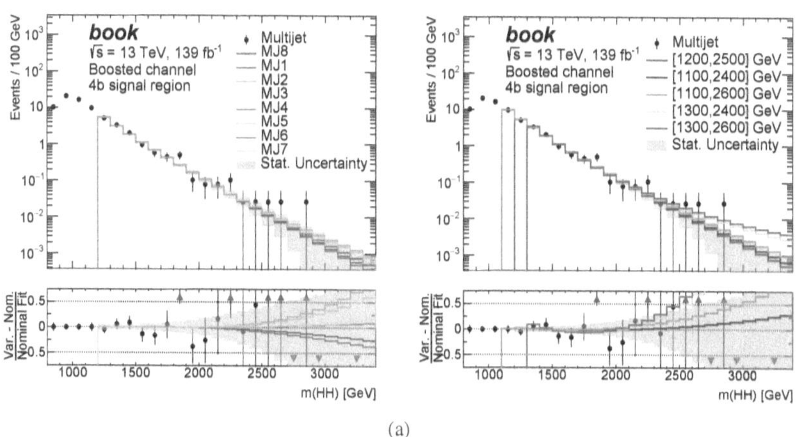

(a)

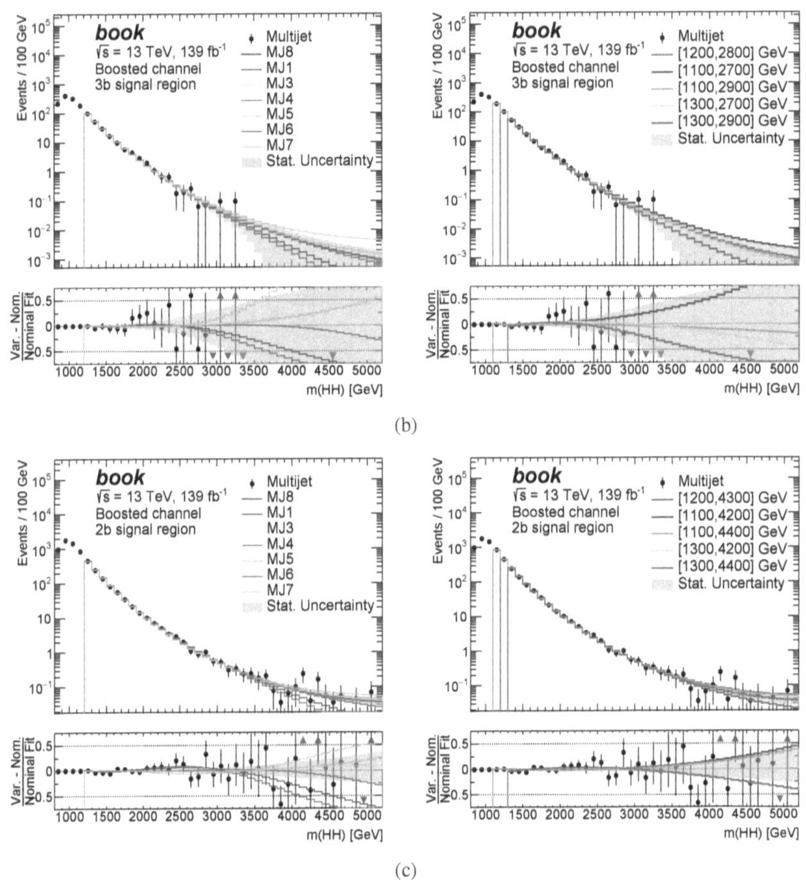

(b)

(c)

Figure 2.20: Result of the smoothing fits on the QCD model in the (a) $4b$, (b) $3b$ and (c) $2b$-split signal regions. The left column shows all dijet functions while the right shows the various choices of fit range. The gray bands show statistical error on the nominal fit.

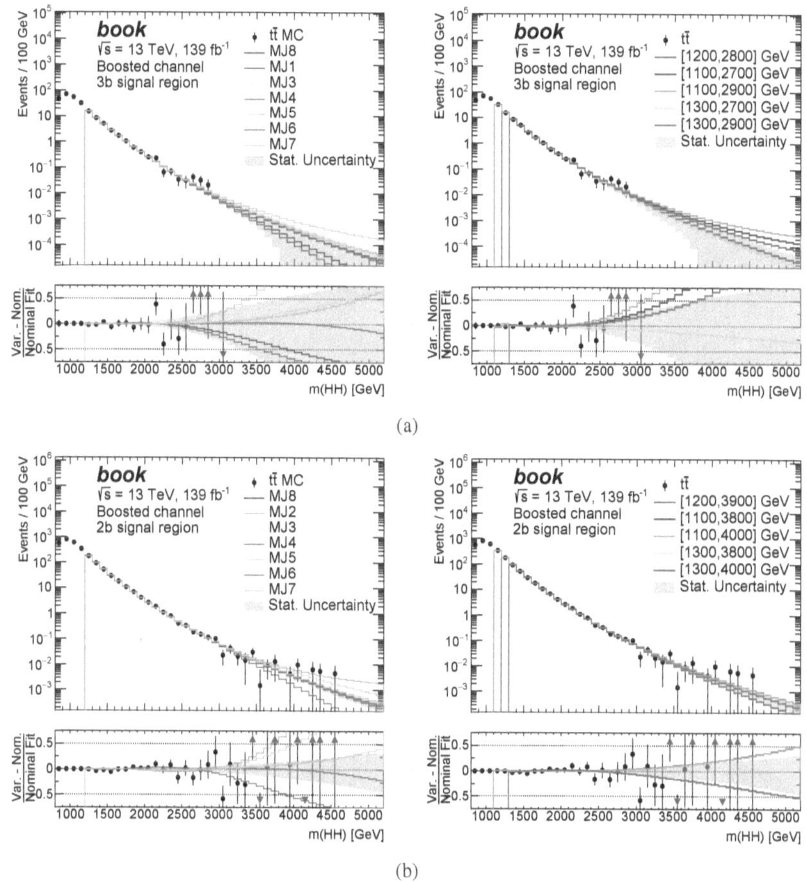

Figure 2.21: Result of the smoothing fits on simulated $t\bar{t}$ in the (a) $3b$ and (b) $2b$-split signal regions. The left column shows all dijet functions while the right shows the various choices of fit range. The gray bands show statistical error on the nominal fit. The shape of the $4b$ $t\bar{t}$ model is taken from the $3b$ region.

2.5.3 Non-closure Uncertainties

Two additional background uncertainties are defined to address any potential biases not covered elsewhere. In particular, they cover differences in shape between the QCD distributions of the signal regions and the corresponding low-tag channels where the QCD models are derived. These differences are estimated in two ways: firstly, using multijet MC and secondly, using validation region data.

For the first method, the background estimation procedure is replicated using only MC simulation. A new reweighting function is derived from the sum of the multijet and $t\bar{t}$ MC samples, using the method described in Section 2.4.1. The reweighted MC is then normalized using the fit method described in Section 2.4.2. In order to define a shape uncertainty, the reweighted low-tag MC prediction is compared to the high-tag MC. An expanded $m_{H_1} - m_{H_2}$ plane region corresponding to the combined signal and validation regions is used to improve statistical precision. Small differences are observed in each region and each such non-closure is fit to a line as shown in Figure 2.22. The observed non-closure is reflected about the nominal background to provide an uncertainty band on the background prediction in the final fit. This method of non-closure estimation assumes that the residual differences between low-tag and high-tag regions, that the reweighting procedure is unable to correct, are similar in the MC simulation and in data. The fact that the observed non-closure is small also serves as a useful validation of the background estimation procedure.

The second method of estimating non-closure uncertainty is done using validation region data. As the validation data is expected to be similar to a background-only signal region, this also serves as a check of the background model. The shape of the QCD model, derived from the low-tag signal region, is compared to that of the high-tag $t\bar{t}$-subtracted validation region data. A non-closure uncertainty is then defined from differences observed in those bins with significant numbers of events. A downside of this method is that statistical fluctuations in the validation region can be quite large and, within uncertainties, the $3b$ and $4b$ shapes were found to agree with the background prediction. The deviations in all three b-tagging channels are shown in Figure 2.23. A difference in the shape of the turn-on curve at low mass is observed in the $2b$-split region and applied as a

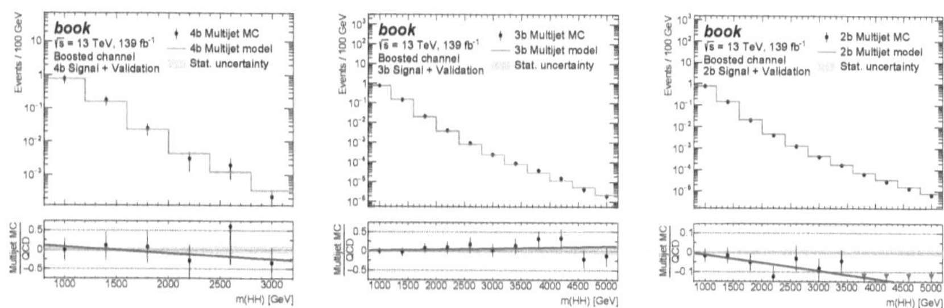

Figure 2.22: Background models constructed from multijet simulation compared to the MC prediction in the (a) $4b$, (b) $3b$ and (c) $2b$-split regions. The comparison is done in a combined signal and validation region, and the bins of the plots are widened, to better show trends in the ratio. The observed non-closure is fit to a line to define a shape uncertainty on the data-driven background.

shape uncertainty in the final fit.

While some bin-by-bin differences can be seen in the $3b$ and $4b$ regions, these appear to be statistical fluctuations so, in practice, a non-closure uncertainty is only applied in the $2b$-split region. Following similar reasoning, the fluctuations in the tail of the $2b$-split distribution are also ignored for the purposes of this systematic. As the uncertainties obtained from the two non-closure estimates have significantly different shapes, they appear to be measuring different ways the background can be mis-modelled. Therefore, both uncertainties are applied in the final fit.

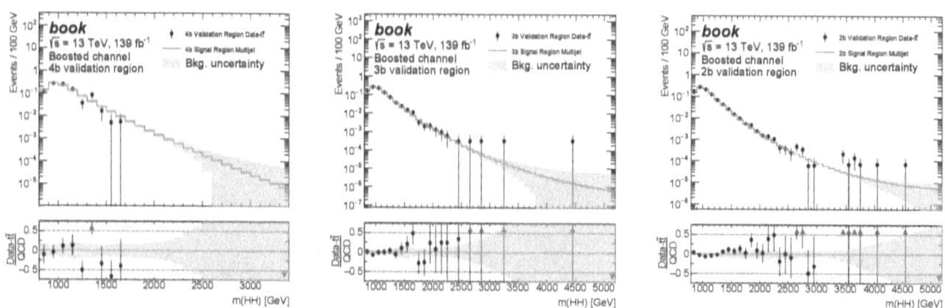

Figure 2.23: Comparisons in the (a) $4b$, (b) $3b$ and (c) $2b$-split validation regions of the shape of the $t\bar{t}$-subtracted data to the multijet background models derived in the corresponding signal regions. The gray band shows the sum of the background modelling uncertainties.

2.5.4 Other Uncertainties

In addition to the uncertainties of the background estimation method, there are a number of other sources of uncertainty considered. These include experimental uncertainties, e.g. those related to the reconstruction of the various physics objects by the detector, and theoretical uncertainties related, e.g. to the model used to simulate the parton shower. Experimental uncertainties are evaluated by dedicated teams within the ATLAS collaboration and incorporated as Bayesian priors in the statistical analysis, as described in Section 2.6.1. The priors associated with theoretical uncertainties, meanwhile, are assessed by varying the assumptions used in event simulation or by parameter changes in the model. All of these uncertainties are generally applied only to simulated samples. While the multijet background is estimated in a data-driven way, it is still sensitive to these uncertainties, through the low-tag $t\bar{t}$ sample used, and the effect of each is considered. A list of all the uncertainties considered, along with brief descriptions, can be found in Table 2.7. Plots of those systematic variations with large impact or the required additional study can be found in Appendix B.

$t\bar{t}$ Simulation

Additional $t\bar{t}$ variations are considered to account for uncertainties in the computational models used for $t\bar{t}$ MC generation. These cover the matrix element calculation, parton shower modelling, renormalization and factorization scales, and the h_{damp} parameter, which tunes the amount of additional hard radiation in the sample and is typically set to $h_{\mathrm{damp}} = 1.5\, m_{\mathrm{top}}$. In addition, uncertainties on the parton distribution functions were investigated and found to be smaller than the statistical uncertainty in the sample.

The matrix element uncertainty is evaluated by comparing aMC@NLO +PYTHIA 8 samples to the nominal POWHEG +PYTHIA 8. Parton shower uncertainties use a comparison POWHEG +HERWIG 7 samples to POWHEG +PYTHIA 8. The differences between samples are then symmetrized to provide an uncertainty band in each of the $4b$, $3b$ and $2b$-split signal regions, as shown in Figure 2.24 for the parton shower uncertainty. Other uncertainties are derived using alternate event weights calculated by PYTHIA 8 when the samples are generated. This is true, for example, for the renormalization and factorization scale uncertainties. For the uncertainty on final state radiation (FSR), the down variation provided ($\mu_R = 0.5$) contains large weights which affect the final distributions. Instead of using this variation, an uncertainty band is constructed by symmetrizing the up variation ($\mu_R = 2.0$). Ultimately this uncertainty has little impact on the fit. For the h_{damp} uncertainty, the variations come from altering parameters in the model of the underlying event. The variation which increases the amount of hard radiation in the event comes from an independent sample generated with $h_{\mathrm{damp}} = 3\, m_{\mathrm{top}}$. The alternate sample available for this variation has lower statistical precision for non-all-hadronic decays than the nominal sample, so instead the prediction from the down variation is symmetrized. Again, this uncertainty has little impact on the fit.

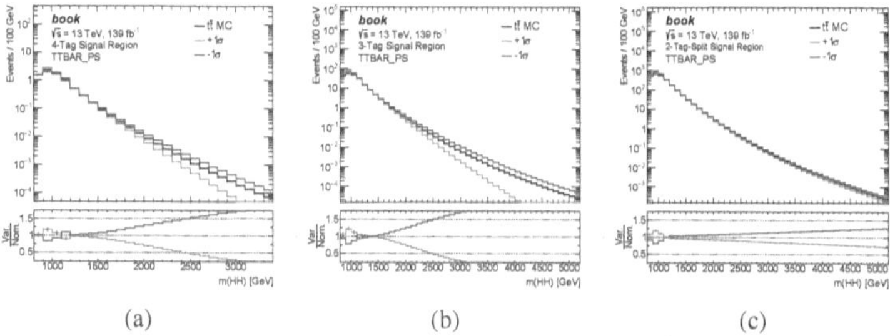

Figure 2.24: Parton shower uncertainty in the (a) $4b$, (b) $3b$ and (c) $2b$-split channels, derived from a comparison of PYTHIA 8 and HERWIG 7 samples. The difference between the two samples is mirrored to create symmetric uncertainty bands. The variations are correlated across channels and controlled by a single nuisance parameter in the likelihood function.

2.5.5 Signal Simulation

The largest source of systematic uncertainty on the generation of the signal samples comes from the modeling of the parton shower. As with the $t\bar{t}$ sample, this uncertainty is evaluated by comparing the PYTHIA 8 and HERWIG 7 generators. The shape of the m_{HH} distribution, and of other kinematic variables, were found to be the same for both generators, so the uncertainty is applied only to the signal normalization. A flat 10% normalization uncertainty is applied in all

b-tagging channels to all signal hypotheses, chosen to be at least as large as the uncertainty seen at any individual mass point.

2.5.6 Summary of Systematics

There are many sources of uncertainty considered in the $HH \rightarrow 4b$ analysis, and each can affect the result in different ways. Table 2.7 contains a full list of uncertainties, along with brief descriptions. Those uncertainties specifically measured for the analysis have already been described in more depth. Table 2.8 summarizes how each group of systematics is implemented in the fit, and the b-tagging channels for which it is used.

Systematic Uncertainty	Brief Description
Background uncertainties are calculated and applied separately in each b-tagging channel. Experimental and theoretical uncertainties are fully correlated between channels.	
Experimental Uncertainties	
Luminosity	Uncertainty on the full Run 2 integrated luminosity, as measured by the LUCID-2 detector [85, 86].
Pileup Reweighting	Uncertainties on pile-up conditions are applied when reweighting simulations to match data.
Jet Energy Scale (JES)	Uncertainty on the reconstruction of large-R jet energies from detector inputs [87, 88]. Applied as 30 independent NPs.
Jet Energy Resolution (JER)	Uncertainty on the precision of jet energy reconstruction [87, 88].
Jet Mass Scale (JMS)	Uncertainty on jet mass reconstruction [87, 88]. Calculated separately from JES and applied as 6 independent NPs.
Jet Mass Resolution (JMR)	Uncertainty on the precision of jet mass reconstruction [87, 88]. Separate NPs used for Higgs boson jets and top quark jets.
DL1r Efficiency	Uncertainty on DL1r tagging efficiencies [46, 50, 49]. 3 NPs are used for b-tagging rates, 4 for c-tagging and 4 for l-tagging.
DL1r SF Extrapolation	Uncertainty due to extrapolation of SFs to track-jets with $p_T > 400\,\text{GeV}$ [46].

Table 2.7: continued on next page

Systematic Uncertainty	Brief Description
Theoretical Uncertainties	
$t\bar{t}$ Matrix Element	Uncertainty in the matrix elements are measured by comparing POWHEG and aMC@NLO predictions.
$t\bar{t}$ Parton Shower	Uncertainty in the parton shower model is measured by comparing PYTHIA 8 and HERWIG 7.
$t\bar{t}$ Hard Radiation	Uncertainty in $t\bar{t}$-associated radiation is assessed by varying the h_{damp} parameter in the model.
$t\bar{t}$ μ_R, μ_F, FSR	Uncertainties in renormalization and factorization scales, and in final state radiation (FSR) are assessed by sample weight variations in PYTHIA 8.
$t\bar{t}$ PDF	Uncertainties on the PDFs are assessed using an ensemble of weight variations in PYTHIA 8.
Signal Parton Shower	Uncertainty in the parton shower model is measured by comparing PYTHIA 8 and HERWIG 7.
Background Estimation Uncertainties	
Normalization Fit	Uncertainty in μ_{QCD} and $\alpha_{t\bar{t}}$ from the fit.
CR Variations	Uncertainty due to choice of CR used for the fit.
Extrapolation	Uncertainty due to extrapolation of μ_{QCD} and $\alpha_{t\bar{t}}$ from CR to SR.
Shape Fit	Uncertainty in the three parameters of the MJ8 function.
Fit Function	Uncertainty due to function used for the fit.
Fit Range	Uncertainty due to range used for the fit.
Non-Closure	Two uncertainties used to assess unknown biases, one defined using multijet MC and the other using VR data.

Table 2.7: Experimental uncertainties considered in the boosted $HH \to 4b$ analysis.

systematic	type	region(s)	corr.	signal	$t\bar{t}$	QCD Model
Luminosity	Norm	all	✓	✓	✓	✓
Jet systematics	Shape & Norm	all	✓	✓	✓	✓
b-tagging systematics	Shape & Norm	all	✓	✓	✓	✓
$t\bar{t}$ simulation	Shape & Norm	all	✓		✓	✓
Signal simulation	Shape & Norm	all	✓	✓		
Background extrapolation	Norm	all				✓
CR Variation	Norm	all			✓	✓
Smoothing systematics	Shape & Norm	all			✓	✓
Non-closure	Shape & Norm	$2b$-split			✓	✓

Table 2.8: Summary of systematics including the type of systematic applied and the samples it applies to. The 'corr.' column indicates whether these NPs are correlated between the different regions. Note that the QCD model is affected by all theory systematics affecting the $n_b - 1$ $t\bar{t}$ MC.

2.6 Statistical Analysis

The statistical analysis in this search is similar to techniques used in previous rounds of this analysis: a profile-likelihood fit is performed simultaneously in the $4b$, $3b$ and $2b$-split channels using the m_{HH} variable as the final discriminant [69]. The relative contributions of each b-tagging channel change for each mass hypothesis and each channel is only included for the range of masses where it contributes significantly. The $4b$ channel, therefore, is only used for $m_{HH} \leq 3\,\text{TeV}$ while the $2b$-split channel is used for $m_{HH} \geq 2\,\text{TeV}$. The $3b$ channel is included for all mass points.

2.6.1 Hypothesis Testing

The statistical test used for hypothesis testing is defined by taking ratios of profiled likelihood functions. The likelihood function for a specific hypothesis can be constructed as follows: given a histogram with entries $\mathbf{n} = (n_1,...,n_N)$, the expectation value of the number of events in each bin can be written as

$$E[n_i] = \mu s_i + b_i, \tag{2.10}$$

where s_i and b_i are the signal and background predictions in bin i, and μ is the signal strength. A value of $\mu = 0$ corresponds to the background-only hypothesis, while $\mu = 1$ is the nominal signal hypothesis. The values s_i and b_i in general depend on a some sets of nuisance parameters, $\boldsymbol{\theta}_s$ and $\boldsymbol{\theta}_b$, that characterize the underlying probability density functions $f_s(m_{HH}; \boldsymbol{\theta}_s)$ and $f_b(m_{HH}; \boldsymbol{\theta}_b)$. These nuisance parameters (NPs) correspond to the systematic uncertainties described in Section 2.5 and their values are constrained by auxiliary measurements incorporated into the likelihood function. The likelihood function takes the following form:

$$\mathcal{L}(\mu, \boldsymbol{\theta}) = \prod_{i=1}^{N} e^{-(\mu s_i + b_i)} \frac{(\mu s_i + b_i)^{n_i}}{n_i!} \prod_k \mathcal{SN}(\theta_k), \tag{2.11}$$

where $\boldsymbol{\theta}$ contains NPs for both signal and background distributions, each assumed to itself follow a two-sided Gaussian (split-normal) distribution, $\mathcal{SN}$. The likelihood function itself describes a high-dimensional surface in parameter-space whose peak corresponds to the model parameters most likely to produce the observed data. The value of the likelihood at this peak is used as a goodness-of-fit metric for the model. Furthermore, the likelihood function is profiled to obtain the NP values, $\theta(\mu)$, that maximize it for a given signal strength, μ. The values of θ that maximize the likelihood for a given signal normalization are labelled $\hat{\hat{\theta}}(\mu)$, while the global maximum of the likelihood function is given by $\mathcal{L}(\hat{\mu}, \hat{\theta}(\mu))$.

Hypothesis testing is performed using the one-sided profile likelihood ratio test statistic $\tilde{q}_\mu$:

$$\tilde{q}_\mu = \begin{cases} -2\ln \frac{\mathcal{L}(\mu,\hat{\hat{\theta}}(\mu))}{\mathcal{L}(0,\hat{\hat{\theta}}(0))} & \hat{\mu} < 0 \\[2ex] -2\ln \frac{\mathcal{L}(\mu,\hat{\hat{\theta}}(\mu))}{\mathcal{L}(\hat{\mu},\hat{\theta}(\hat{\mu}))} & 0 \leq \hat{\mu} < \mu \\[2ex] 0 & \mu < \hat{\mu} \end{cases} \tag{2.12}$$

$\tilde{q}_\mu$ tests the compatibility of the data with two competing statistical models: the model found by maximizing the likelihood, and model in which the signal strength is constrained to a particular value μ. A small value of $\tilde{q}_\mu$ indicates the data is equally consistent with both models, i.e. that the constraint applied is supported by observation. The chosen test statistic both ensures that an upwards fluctuation of the signal does not serve as evidence against the signal, ($\mu < \hat{\mu}$ case) and that a downward fluctuation of the background is not evidence against the background ($\hat{\mu} < 0$ case).

In order to quantify compatibility of the observed test statistic with a particular hypothesis, one needs to measure the distribution of the test statistic under that hypothesis. While these distributions are difficult to measure in general, for large datasets and assuming a Gaussian distribution of $\hat{\mu}$, the profile likelihood ratio approaches a non-central χ^2 distribution. In this asymptotic limit, the distribution is fully characterized by the mean and standard deviation of $\hat{\mu}$, which can be measured from a single dataset using the method described in Ref. [89]. The Asimov dataset used in this method is defined such that the likelihood is maximized when all parameters take on their nominal values. In practice, the asymptotic method works moderately well even in cases with few events per bin. For the high mass signals, where the limit depends on bins with no observed events, an ensemble Monte Carlo method is used to test the asymptotic limit. The so-called toy method involves randomly generating large numbers pseudo-experiments or toys. For each toy a pseudo-data distribution is generated from the probability distribution function of either the background-only or signal+background hypothesis, the test statistic is then measured, and the distributions of the test statistic from all toys is used to set limits. Due to the computation time required to generate and evaluate large numbers of toys, this method was only used for the high mass signals, and only a small grid of signal strengths were tested for each mass hypothesis. Limits computed with the toy method were found to agree well with those from the asymptotic method for the spin-2 signal, but not for the narrower spin-0 resonances. In the final result the asymptotic method is used for signal masses up to 3 TeV and the toy method is used for higher masses. More details on the toy method can be found in Appendix C.

With the test statistic distributions calculated, it is possible to quantify the level of incompatibility between the data and the background-only hypothesis. This quantity, the p-value of the null hypothesis, is defined as

$$p_0 = \int_{q_{\text{obs}}}^{\infty} f(q|H_0)dq,\tag{2.13}$$

i.e. the probability of observing a q value greater than or equal to q_{obs} under the null hypothesis. The p-value is often expressed as a significance, Z, defined as

$$Z = \Phi^{-1}(1 - p),\tag{2.14}$$

where Φ is the cumulative distribution of the unit Gaussian. In high-energy physics, a significance of 5σ is required to claim discovery, corresponding to a p_0 value of 2.87×10^{-7}.

Due to the nature of the likelihood-ratio test, the value of p_0 depends explicitly on a specific signal hypothesis that enters through the global maximum of the likelihood function. One can think of this ratio as restricting the statistical test to a subset of the mass range corresponding to the signal hypothesis being tested. Each signal hypothesis therefore requires a separate calculation of the discovery significance looking for the evidence for discovery of a new particle at that specific mass. The significance of the test for a single signal mass is refered to as *local* significance. In the boosted $HH \rightarrow 4b$ analysis alone more than a dozen signal hypotheses are tested for each signal model, with a similar number of tests in the resolved analysis. A *global* significance must therefore be calculated to account for the increased probability of finding large local significances when making many measurements. Global significances are calculated using an ensemble method in which toys are randomly generated following the background-only hypothesis. The maximum local significance, among all signal masses, of each toy is calculated to obtain a distribution of Z_{local} values. p_{global} is then defined as

$$p_{\text{global}} = \int_{Z_{\text{ref}}}^{\infty} f(Z_{\text{local}}|H_0)dZ, \tag{2.15}$$

where Z_{ref} is a particular reference Z_{local} used. Due to the computation time required to calculate the global significance, this was only done for the combined results in Section 2.8.

2.6.2 Limit-setting Procedure

When calculating limits, the hypothesis test is inverted so that the signal+background hypothesis takes the role of H_0. The value of the signal strength, μ, is varied until a value is found that results in $\text{CL}_s = 0.05$. The CL_s for the test statistic q is defined as:

$$\text{CL}_s = \frac{p_0}{1 - p_1} = \frac{P_{\text{s+b}}(q \geq q_{\text{obs}})}{P_b(q \geq q_{\text{obs}})} \tag{2.16}$$

where $P_{\text{s+b}}(q \geq q_{\text{obs}})$ is the probability of the signal+background model to produce equal or better agreement to the data than observed, and $P_b(q \geq q_{\text{obs}})$ is the probability of the background only model to produce equal or better agreement to the data than observed. Signal strengths greater that which gives $\text{CL}_s = 0.05$ are considered incompatible with the data and excluded at 95% confidence level. This signal strength can then be converted to a limit on the cross section $\sigma(X \rightarrow hh \rightarrow b\bar{b}b\bar{b})$.

2.6.3 Signal Injection Tests

Signal injection tests were performed to test the ability of the fit to correctly measure signal strength. These tests were performed on Asimov datasets composed of the nominal background and a known number of signal events. The fit was able to reproduce the expected signal strengths for positive signals for all signal masses. Histograms are required to have non-negative counts in each bin, limiting the size of negative signal strengths. This behaviour can be seen in Figure 2.25, which shows the result of the test on the 2 TeV spin-0 signal hypothesis.

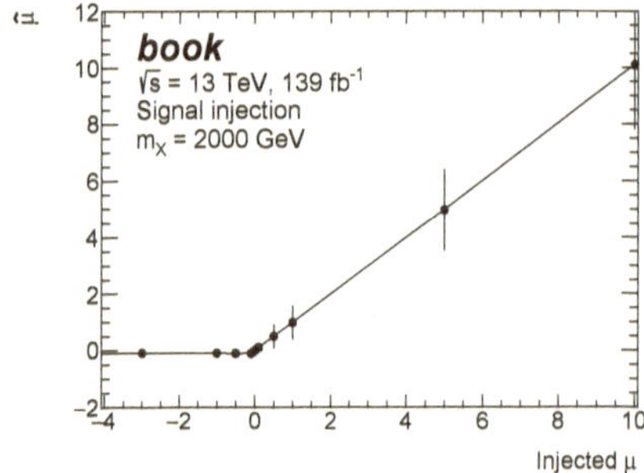

Figure 2.25: Result of signal injection tests with 2 TeV spin-0 signal. The fitted signal matches expectation for positive signal strengths.

2.6.4 Signal Morphing

Additional scalar signal masses are generated by interpolating between those generated by the full ATLAS simulation. The interpolated masses are m(X) = 2250, 2750, 3500, 4500 GeV, corresponding to the additional masses generated for the spin-2 model but not the spin-0 model. A linear moment morphing procedure is used to interpolate between the simulated masses immediately above and below the target mass. The normalization of the all the simulated scalar masses are fit to a cubic spline, which is then used to set the normalization for the morphed histograms. To validate the method, the 2500 GeV point is generated by interpolation and compared to the actual simulated sample distribution, as shown in Figure 2.26. The morphing provides an approximation of the interpolated signal sufficient to fill in additional detail in the 95% CL$_s$ limits. The same procedure is used to generate systematic variations for the interpolated mass points.

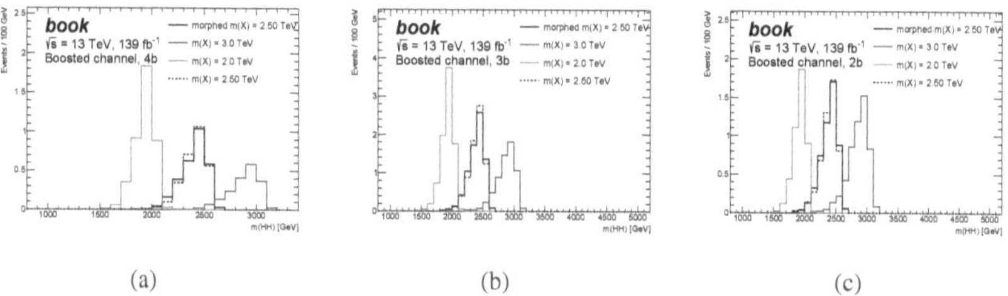

(a)   (b)   (c)

Figure 2.26: Result of using moment morphing to interpolate the 2500 GeV spin-0 signal mass in the (a) 4*b*, (b) 3*b*, and (c) 2*b*-split channels. The 2000 GeV and 3000 GeV mass histograms are used to parametrize the signal. The 2500 GeV spin-0 signal generated from MC is shown for comparison.

2.6.5 Impact of Systematic Uncertainties

The relative importance of the various sources of systematic uncertainty are measured individually and in groups. The maximized likelihood function, $\mathcal{L}(\hat{\mu}, \hat{\theta}(\hat{\mu}))$, is used to measure the individual impact of each nuisance parameter. First, the correlation matrix and the Bayesian posterior distributions of the NPs are extracted from the maximum likelihood fit. Then, for each NP, the likelihood is maximized with the value of that NP fixed to the $\pm 1\sigma$ values of the posterior distribution. The result is a measurement of the impact of each individual NP on $\hat{\mu}$. The posterior distribution and impact are shown together in Figure 2.27 for the 2 TeV signal masses. The points, and error bars, show the mean and width of the posterior probability relative to the prior on the bottom axis, while the color bars show the impact on $\hat{\mu}$ on the top axis. The difference between prior and posterior uncertainties is referred to as the constraint on the NP from the fit, and is generally expected to be small. In addition, the difference between the best-fit value of the NP and the prior value of zero is expected to be generally less than 1σ. Indeed, this is observed in almost all cases. This method of measuring the impact of individual NPs does not take correlations between NPs into account. Moderate correlation between parameters is observed, however, as shown in

84

Figure 2.27c.

The impact of each source of uncertainty is also assessed on the 95% CL_s limit directly. To do this, NPs are grouped by source and all NPs in a group are fixed to the maximum-likelihood value, $\hat{\theta}(\hat{\mu})$. The limit is then recalculated and the relative difference between conditional and unconditional limits is measured. The impact of the main sources of uncertainty given in Tables 2.9 and 2.10 for spin-0 and spin-2 models respectively. The relative importance of different sources of uncertainty varies with signal mass and with the relative importance of the different b-tagging channels.

Uncertainty category	Relative impact (%)				
	1000 GeV	1600 GeV	2000 GeV	3000 GeV	5000 GeV
Background m_{HH} shape	21	1.3	0.6	1.2	1.0
Jet momentum/mass scale	0.1	1.5	0.8	4.7	0.5
Jet momentum/mass resolution	4.4	7.4	16	9.5	6.5
b-tagging calibration	0.6	1.8	3.0	0.8	6.3
Theory (signal)	1.8	1.6	1.2	1.7	1.1
Theory ($t\bar{t}$ background)	5.6	0.7	0.5	0.8	0.2
All systematic uncertainties	35	13	21	15	15

Table 2.9: Impacts of the main sources of systematic uncertainty on the expected spin-0 95% CL_s limits. These are defined as the relative decrease in the limit when each set of nuisance parameters is held fixed to its best-fit value instead of being assigned an uncertainty.

Uncertainty category	Relative impact (%)				
	1000 GeV	1600 GeV	2000 GeV	3000 GeV	5000 GeV
Background m_{HH} shape	32	1.9	1.3	2.1	2.0
Jet momentum/mass scale	0.0	2.4	1.6	6.0	0.9
Jet momentum/mass resolution	5.8	8.8	15	9.0	6.7
b-tagging calibration	0.8	1.7	2.8	1.2	5.7
Theory (signal)	1.8	1.5	1.5	2.0	1.1
Theory ($t\bar{t}$ background)	5.3	0.9	1.1	1.3	0.6
All systematic uncertainties	47	16	22	16	15

Table 2.10: Impacts of the main sources of systematic uncertainty on the expected spin-2 95% CL_s limits. These are defined as the relative decrease in the limit when each set of nuisance parameters is held fixed to its best-fit value instead of being assigned an uncertainty.

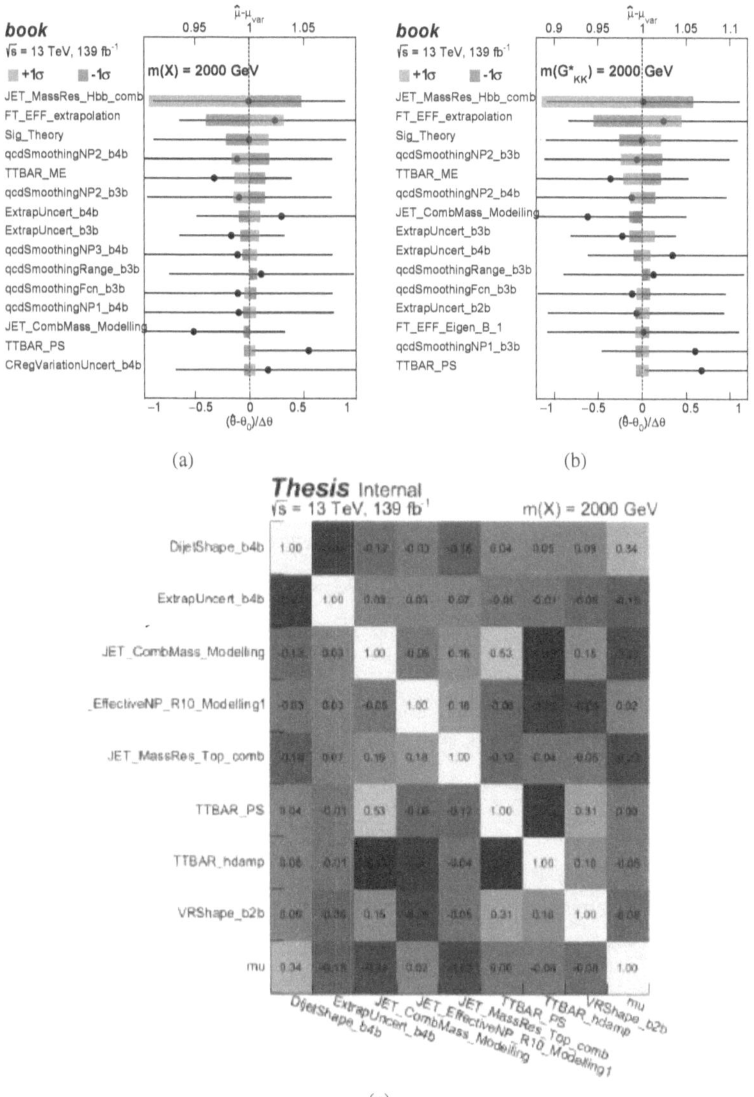

Figure 2.27: Pulls (points, bottom axis) and impacts (bars, top axis) of NPs in the fit to the (a) 2 TeV spin-0 and (b) 2 TeV spin-2 signal hypotheses, as well as the correlation matrix (c) from the fit to the 2 TeV spin-0 signal. Pulls are only shown for the 15 NPs with the highest impact on $\hat{\mu}$ and correlations are only shown for NPs with at least 20% (anti-)correlation to another parameter.

2.7 Results

The primary results of the statistical analysis are the p_0 values and the derived 95% CL_s limits on the signal mass hypotheses. Both are presented in this section, along with the results of the likelihood maximization, namely the best-fit m_{HH} spectrum and values of the nuisance parameters. The best-fit results are used to assess the ability of the model to accurately fit the data. Several checks of the validity of the results are also presented.

2.7.1 Post-fit Distributions

Figure 2.28 shows the best-fit m_{HH} distributions under the background-only hypothesis. That is to say, the distribution with all NPs set to the values $\hat{\theta}(0)$ that maximize the likelihood under the condition $\mu = 0$. In general, the global minima of the likelihood function has a non-zero μ value and depends on the signal hypothesis being tested. Several spin-0 signal hypotheses are overlaid on the plot, normalized to the expected limits, to give an indication of what a detected signal may look like. Figure 2.29 shows the same m_{HH} distribution with spin-2 signals overlaid, illustrating the difference between the two signal models.

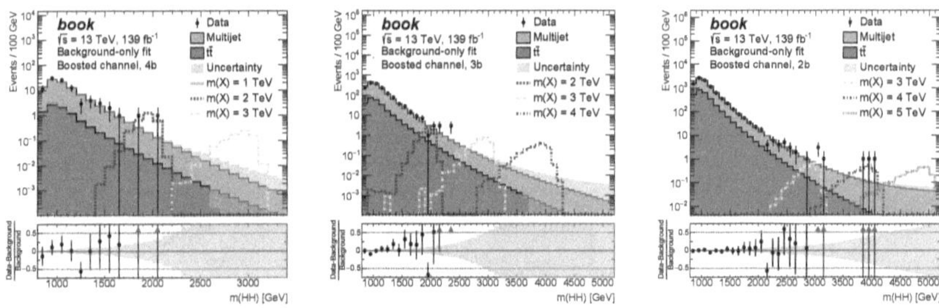

Figure 2.28: Post-fit m_{HH} distributions in the (a) $4b$, (b) $3b$ and (c) $2b$-split channels under the background-only hypothesis. The gray band shows the total post-fit uncertainty on the background model. Representative spin-0 signals are shown normalized to the observed 95% CL_s limits.

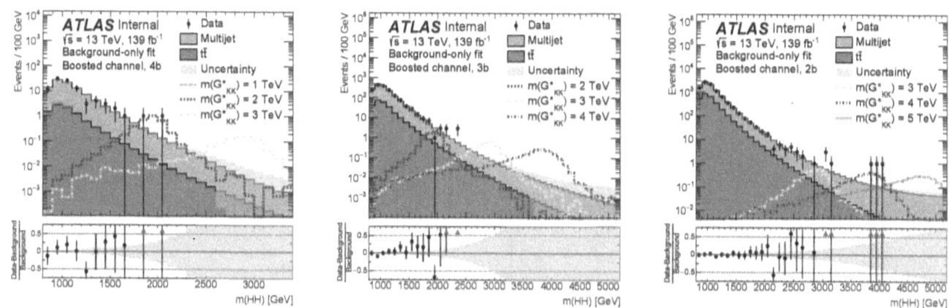

Figure 2.29: Post-fit m_{HH} distributions in the (a) $4b$, (b) $3b$ and (c) $2b$-split channels under the background-only hypothesis. The gray band shows the total post-fit uncertainty on the background model. Representative spin-0 signals are shown normalized to the observed 95% CL_s limits.

2.7.2 Discovery Signficance

Figure 2.30 shows the observed p_0 values at each mass point under the RS and 2HDM models, calculated using the full Run 2 dataset. Each p_0 value measures whether a background-only model, $\mu = 0$, can fit the data as well as the signal+background model that maximizes the likelihood function, $\mu = \hat{\mu}$, for that particular signal hypothesis. The largest local significance is at 4 TeV with $Z_{\text{local}} = 1.85\,(1.41)$ for the spin-0 (spin-2) model. The large significance here comes from the observation of three events in the $2b$-split channel, visible in Figure 2.28 and Figure 2.29, where the background predicts fewer than one event per bin. These few points provide much greater evidence for the spin-0 model, which predicts a narrow peak, than for the spin-2 model, which predicts a peaked but quite broad enhancement to the HH cross-section. Even so, neither local significance value rises to the level of discovery. The true significance is lower still because the look-elsewhere effect has not been taken into account in these plots. Since no discovery is made, limits are set on the allowed cross-sections of the spin-0 and spin-2 signal models.

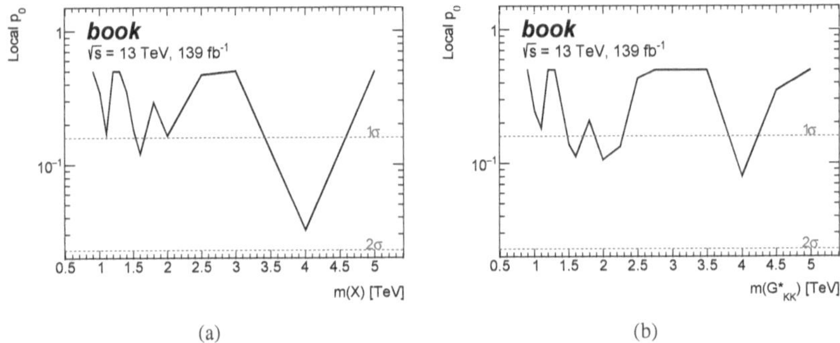

Figure 2.30: Local p-value comparing the consistenty of the $\mu = 0$ and $\mu = \hat{\mu}$ hypotheses with the data. P-values are calculated separately for each of the (a) spin-0 and (b) spin-2 signal masses. A p-value of 2.87×10^{-7}, corresponding to a local significance of 5σ is required to claim discovery.

2.7.3 Expected and Observed Asymptotic Limits

Figure 2.31 shows the expected and observed limits on the RS and 2HDM models using the full Run 2 dataset. The theoretical prediction for the RS model is taken from Ref. [8]. The observed limit, drawn as a solid black line, corresponds to the minimum value of μ, for each signal hypothesis, that is incompatible with the data. The phase-space above this line is considered to be excluded to the 95% confidence level. As with the discovery significance, small excesses in the data above the background-only hypothesis push the limits to higher values at specific masses. Here though, small deficits in the data can also serve as evidence against a signal and push the limits to lower values. The expected limit, and the 1σ and 2σ uncertainty bands, are derived from Asimov datasets and show essentially how the limits would have appeared if the data had matched the background prediction exactly, or a 1σ or 2σ variation of the background. The cumulative effect of small excesses, particularly in the $3b$ channel, push the observed limits above the expected limits over much of the mass range, although the difference between the two is within the 1σ band. Differences between the limits for the spin-0 and spin-2 models arise primarily because of the difference between the widths of the resonances.

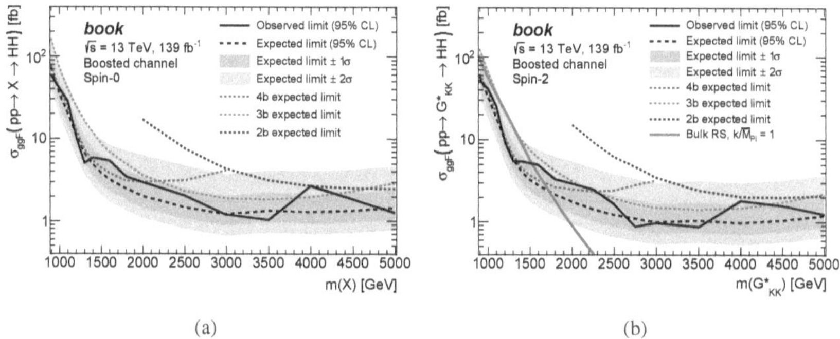

(a)   (b)

Figure 2.31: Expected and observed limits from the boosted $HH \rightarrow 4b$ analysis on (a) spin-0 and (b) spin-2 signal models derived using the full Run 2 ATLAS dataset of 139 fb^{-1} of $\sqrt{s} = 13$ TeV proton–proton collision data. The contributions of the individual $4b$, $3b$ and $2b$-split channels are shown in green, pink, and blue respectively.

2.7.4 Expected and Observed Toy Limits

Toy limits are calculated for the signal mass above 3 TeV, where few background events are predicted. The toy method is described in detail in Appendix C and is used to relax certain assumptions made in the asymptotic method. Figure 2.32 shows a comparison between the expected and observed toy limits and the asymptotic limits, for both the RS and 2HDM signal models. The toy limits are generally consistent with the asymptotic limits, in particular for the spin-2 model. The limits for the spin-0 model diverge slightly as the signal mass increases, as expected, resulting in a 20% difference at 5 TeV. Large differences are also observed in the -2σ error band. This comes from the data being limited to at least zero events per bin, an effect not taken into account in the asymptotic calucation. The toy limit, being more correct, is used in the final result.

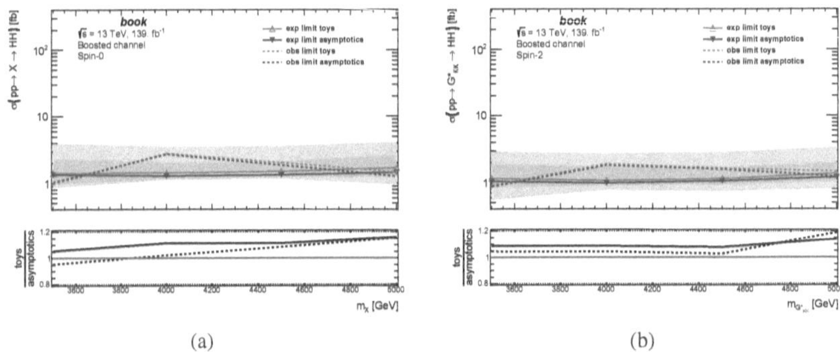

(a)   (b)

Figure 2.32: Comparison of the toy and asymptotic limits from the boosted $HH \rightarrow 4b$ analysis on (a) spin-0 and (b) spin-2 signal models derived using the full Run 2 ATLAS dataset of 139 fb^{-1} of $\sqrt{s}$ = 13 TeV proton–proton collision data. The results are mostly consistent but the toy limits diverge from the asymptotic values by up to 20% at high masses due to a breakdown of the assumptions used in the asymptotic calculation. The error bands shown here are also calculated using the toy method.

2.8 Statistical Combination

The final results of the $HH \rightarrow 4b$ analysis combine the resolved and boosted channels to compute limits across the full mass range of 251-5000 GeV. The datasets used by the resolved and boosted channels are entirely orthogonal, which allows for a simple statistical combination by taking the product of the individual likelihoods. As the two channels use different jet definitions, b-tagging algorithms and background estimation methods, almost all systematics are uncorrelated between them. The only exceptions are the uncertainty on the ATLAS luminosity measurement and the uncertainties in the signal MC generation, which are fully correlated between the channels.

An excess above 2σ in local significance is observed in the combined limits for the 1100 GeV mass point, and excesses above 1.5σ are observed at 1400, 1500 and 4000 GeV. The observed local significances for the 1100 GeV mass point are 2.5σ for both spin-0 and spin-2 signal hypotheses. The global significances are 0.9 and 1.4 for spin-0 and spin-2 models respectively. For the 1400 GeV mass point, the combined significances are 1.6 and 1.7σ for the spin-0 and spin-2 signal hypotheses. We therefore conclude that no evidence for either spin-0 or spin-2 signal model

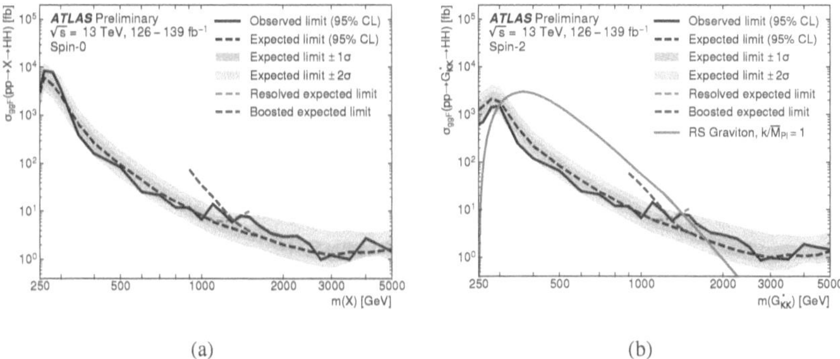

(a)   (b)

Figure 2.33: Expected and observed 95% CL upper limits on the cross section times branching ratio of resonant $X \to HH$ production in the spin-0 and a spin-2 signal models. The toy method is used to derived limits for signal masses above 3 TeV. The theoretical prediction for the RS model is also shown.

is present in the ATLAS Run 2 dataset. Instead, we set limits on the cross-section of Higgs boson pair production assuming the SM branching ratio to $b\bar{b}b\bar{b}$ of 58%, as shown in Figure 2.33. For signal masses up to and including 3 TeV, the asymptotic method is used to derive limits, but at higher masses the toy method is used instead. The theoretical prediction for the RS model with $k/\overline{M}_{\mathrm{Pl}} = 1$ is taken from Ref. [8]. While we expected to exclude this model in the mass range from 304 GeV to 1730 GeV, in the final result the model is excluded for masses between 298 GeV and 1440 GeV.

Chapter 3: $g \rightarrow bb$ Calibration

3.1 Calibration Overview

The $g \rightarrow bb$ calibration presented here aims to calibrate a novel double-b-tagger for use in ATLAS analyses. b-tagging algorithms, as described in Section 1.10, are widely used in ATLAS particularly for analyses focused on Higgs boson and top quark decays. These algorithms, including the DL1r algorithm [42] used in the $HH \rightarrow 4b$ analysis, classify jets based on the heaviest flavor hadron they are likely to contain, b, c, and light. Recently, the Xbb2020 algorithm has been developed specifically to identify double-b decays fully contained within large-R jets [90]. This algorithm classifies a large-R jet as a coming from Higgs boson, top quark or multijet process, based on the b-tagging scores of the associated variable-radius track-jets. The Xbb2020 algorithm shows large improvements in classification performance by taking into account correlations between the two b-hadrons in Higgs boson decays. Just as with DL1r, a set of working points is defined based on tagging efficiency on simulated $H \rightarrow bb$ decays. These working points must be calibrated before the algorithm can be used in analyses and the calibration method used for DL1r, which uses isolated b-jets from top quark decays, cannot be used. Several new calibration methods are currently under development within ATLAS, two using $Z \rightarrow bb$ decays and one using $g \rightarrow bb$ decays, presented here.

A calibration, in this context, quantifies the difference in some observable(s) between the measurement in real and simulated data with the goal of defining a procedure to correct inaccuracies in the simulation. After that procedure is applied to the simulation, the targetted observable(s) should match the data exactly. While it may seem natural to correct the distribution of Xbb2020 classification scores, or the final discriminant, this has proven prohibitively difficult in other ML-based algorithms, as the scores have a complex dependence on many input variables. Instead one can correct the efficiency of a small set of cut values on the final discriminant, i.e. the working points. These working points are chosen to be broadly applicable to a wide range of analyses, though they are not fully optimized for any. The calibration aims to provide a set of scale factors (SFs) that can be used to adjust the efficiency seen in simulation to match that of data, as well as to quantify the uncertainties associated with each SF. For b-tagging algorithms, separate calibrations are used to correct the b-tagging efficiency and the mis-tag rates of charm and light jets. The $g \rightarrow bb$ calibration presented here aims only to correct the tagging efficiency on Higgs bosons decaying to two b-hadrons.

Scale factors are defined by the following equation,

$$SF = \frac{\epsilon_{\text{data}}}{\epsilon_{\text{MC}}} = \frac{N_{\text{data}}^{tagged}/N_{\text{data}}}{N_{\text{MC}}^{tagged}/N_{\text{MC}}}, \tag{3.1}$$

where N_{data} is the number of true $b\bar{b}$ decays in data and N_{MC} is the number in Monte Carlo. The measurement of the true $b\bar{b}$ fraction of the data sample forms the crux of the measurement.

This calibration exploits the topological similarities between $g \rightarrow bb$ and $H \rightarrow bb$ decays to perform a calibration on a dataset of multijet events, independent of the $H \rightarrow bb$ signal samples that may be used by analyses. The Xbb2020 algorithm is described in Sec. 3.2, as well as the differences between $g \rightarrow bb$ and $H \rightarrow bb$ jets. The event selection used in the $g \rightarrow bb$ calibration is described in Sec. 3.3, while Sec. 3.4 describes the profile likelihood fit used to extract the scale factors. Sec. 3.5 describes the treatment of systematic uncertainties in the measurements. The results are presented in Sec. 3.6.

3.2 Double-b-tagging

The long lifetime of b-hadrons is exploited in several different ways to create the b-tagging algorithms described in Section 1.10 and used in the $HH \rightarrow 4b$ analysis. These taggers are limited, however, by only considering a single b-hadron decay at a time. Taggers that consider only a single track-jet cannot take correlations between the two b-quarks of a Higgs boson decay into account.

The Xbb2020 algorithm calibrated here is optimized specifically for identifying high-p_T $H \to bb$ decays. A large-R jet is used to capture the entire Higgs decay, while smaller variable-radius track-jets are used to resolve the individual b-jets. Xbb2020 uses a DNN to process the DL1r scores of up to three subjets along with the basic kinematics of the large-R jet. It produces three classification scores for each large-R jet to differentiate between Higgs, top, and multijet processes [90]. As with DL1r, a final discriminant is given by the ratio of the individual scores:

$$D_{\mathrm{Xbb}} = \ln\left(\frac{p_{\mathrm{Higgs}}}{f_{\mathrm{top}}p_{\mathrm{top}} + (1 - f_{\mathrm{top}})p_{\mathrm{multijet}}}\right), \tag{3.2}$$

where p_{Higgs}, p_{top}, and p_{multijet} represent the Higgs bosons, top quark and multijet scores respectively, and f_{top} is the fraction of the sample coming from top quark decays. For comparison, one can also identify $H \to bb$ decays by requiring two subjets tagged by DL1r or MV2. Fig. 3.1 shows the score distributions of the Xbb2020 and double-DL1r methods on simulated Higgs boson, top quark and multijet samples. The D_{Xbb} distribution shows clear separation between the three cases, unlike the DL1r method. Fig. 3.2 shows the multijet and top quark rejection rates as a function of the Higgs jet efficiency. For any given efficiency, the Xbb2020 algorithm is better able to reject the top quark and multijet backgrounds, indicating that correlations between the subjets provide useful information for the tagging algorithm.

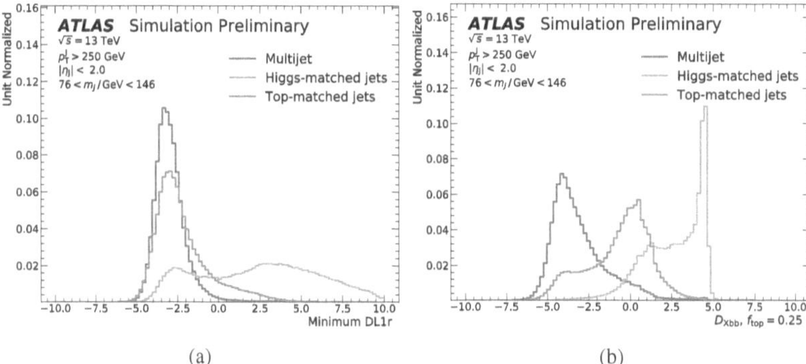

(a)   (b)

Figure 3.1: The double-b-tagging discriminant distributions defined as (a) the minimum DL1r discriminant of the two leading track-jets, and (b) D_{Xbb} with a top quark fraction of $f_{\mathrm{top}} = 0.25$ [90].

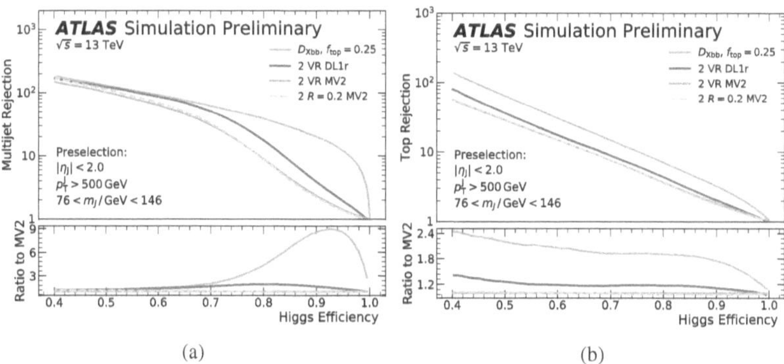

(a)   (b)

Figure 3.2: Multijet (a) and top quark (b) rejection, defined as the inverse of cut efficiency, are compared as a function of $H \rightarrow bb$ tagging efficiency. Tagging is done by either the Xbb2020 algorithm, or by requiring two single-b tags from the DL1r or MV2 algorithms. Separate versions of the MV2 algorithm trained either on variable-radius (VR) track-jets or on R = 0.2 track-jets are considered [90].

3.3 Event Selection

Object Cuts   Extra selection cuts are applied to some physics objects beyond those described in Sec. 1.9. In particular, the calibration procedure uses a set of tracks that is similar to, but does not exactly match the set directly clustered into the variable-radius track-jets. The tracks used to train b-tagging algorithms are those that fall within a ΔR cone whose size decreases with jet p_T. This b-tagging association cone has a width of 0.45 for jet $p_T = 20\,\mathrm{GeV}$ and narrows to 0.26 for jet $p_T = 150\,\mathrm{GeV}$. In case a track is matched to multiple jets, it is associated with the nearest jet [91]. The same track matching procedure is used to determine the inputs for the $g \rightarrow bb$ calibration. Matched tracks are then required to pass the 'loose' criteria defined in Ref. [92] and have $p_T > 0.5\,\mathrm{GeV}$. To reject tracks from pileup vertices, additional requirements on the transverse and longitudinal impact parameters, of $|d_0| < 5\,\mathrm{mm}$ and $|z_0 \sin\theta| < 3\,\mathrm{mm}$ respectively, are applied. These requirements are much looser than the selection generally applied in ATLAS analyses in order to keep as many of the high impact parameter tracks from b-decays as possible. Each track-jet is required to be matched to at least three tracks in order to be used for the calibration.

The muons used for the calibration are required to statisfy the 'loose' identification criteria described in Sec. 1.9, i.e. $p_T > 10\,\mathrm{GeV}$ and $|\eta| < 2.4$. To ensure the muon comes from the primary vertex (or a nearby b-hadron decay), muons are required to have $|d_0| < 2\,\mathrm{mm}$ and $|z_0 \sin\theta| < 2\,\mathrm{mm}$. Muons are matched to the closest track-jet in the $\eta - \phi$ plane, if they fall within the radius of that jet.

Trigger   The $g \rightarrow bb$ calibration uses a logical OR of a set of small-R jet triggers to select events. The efficiency of the trigger is measured as a function of offline small-R jet p_T and each trigger is only used if the event contains a small-R jet for which that trigger would have $> 99\%$ chance of firing. The jet p_T thresholds change every year as LHC run conditions change in order to maintain the same data-recording rate. Table 3.1 lists the HLT and offline p_T thresholds for each trigger used. Data from 2015 uses the same p_T thresholds as 2016, except without the $380\,\mathrm{GeV}$ trigger.

Each trigger below the highest p_T trigger in a given year is *prescaled*. Prescaled triggers do not

Offline Jet p_T	Online Jet p_T			
	2015	2016	2017	2018
420 GeV	-	-	463.5 GeV	454.5 GeV
380 GeV	-	436 GeV	-	-
360 GeV	416.5 GeV	416.5 GeV	395.5 GeV	388.5 GeV
260 GeV	297.5 GeV	297.5 GeV	282.5 GeV	282.5 GeV
175 GeV	205.5 GeV	205.5 GeV	193.5 GeV	193.5 GeV

Table 3.1: Online and offline p_T thresholds by data-taking year for each trigger used.

always fire when the trigger conditions are passed. Instead, they fire on every P-th event passing the condition, where P, the prescale, is set to keep the trigger rate at a manageable level. Each event passing a prescaled trigger therefore represents P total events that could have passed. When used in the calibration, each data event is weighted by the combined prescale of all triggers for which it passes the online jet p_T threshold.

Kinematic Cuts Events are selected if they contain a valid $g \rightarrow bb$ candidate jet. Large-R jets are used to fully contain the $g \rightarrow bb$ decay, and track-jets are used to reconstruct the individual b-hadrons. A valid $g \rightarrow bb$ candidate is defined as a large-R jet containing at least two variable-radius track-jets. In order to select semi-leptonic b-hadron decays, at least one of the track-jets in the $g \rightarrow bb$ candidate is required to contain a muon. In the simulated multijet sample, this requirement increases the fraction of heavy-flavor events approximately threefold. The highest p_T track-jet containing a muon is referred to as the 'muon-jet' while the other track-jets associated to the $g \rightarrow bb$ candidate are referred to as 'non-muon-jets'. If multiple $g \rightarrow bb$ candidates are present in an event then only the highest p_T one is used.

As described above, a set of cuts are applied to avoid the trigger biasing results. The event is required to have a small-R jet matched to the online jet that fired the trigger (within $R = 0.4$), and the $g \rightarrow bb$ candidate is required to be on the opposite side of the event ($R > 1.5$).

Events are binned in the p_T of the large-R jet and separate scale factors are derived for each bin. The bins are [250, 350, 400, 450, 500, 550, 600, 750, 1000] GeV. Wider bins are used at low p_T, where large trigger prescales reduce the effective statistical precision of the sample, and at high

Label	Category definition
BB	At least two track-jets contain *b*-hadrons
BL	Exactly one track-jet contains *b*-hadrons
CC	No *b*-hadrons and at least two track-jets contain *c*-hadrons
CL	No *b*-hadrons and exactly one track-jets contains *c*-hadrons
LL	No track-jets contain *b*- or *c*-hadrons

Table 3.2: Flavor category labels and definitions used in the $g \rightarrow bb$ calibration.

p_{T} where less data is available.

3.3.1 Collinear Track-jet Veto

As mentioned in Section 2.3.7, in some cases a variable-radius track-jet can be fully contained inside another. These events are not used when training *b*-tagging algorithms and are vetoed from the $g \rightarrow bb$ calibration to avoid potential biases. The veto is only applied if overlapping track-jets are both matched to a valid $g \rightarrow bb$ candidate passing all other selection criteria.

3.3.2 Flavor Categorization

Events in the simulation are assigned flavor categories based on the quarks contained in the $g \rightarrow bb$ candidate. Each track-jet is labelled according to the heaviest simulated quark it contains: *b*, *c*, or light. $g \rightarrow bb$ candidates are then categorized according to the two heaviest flavor labels among up to three of the associated track-jets. The muon-jet is always considered in this categorization, and then the highest p_{T} track-jets among the rest. The final set of categories are summarized in Table 3.2.

The D_{Xbb} distribution with a top fraction of $f_{\mathrm{top}} = 0.25$ is shown in Figure 3.3 for each flavor category and for data. As expected, jets in the *BB* category have scores similar to those of true $H \rightarrow bb$ decays. Jets in the *BL* category behave similar to those from hadronic top decays (which generally contain a single *b*-jet). *c*-jets can contain displaced vertices but often do not, resulting in a broad D_{Xbb} distribution for *CC* and *CL* categories between the *BL* and *LL* peaks. While only two labels are used to categorize each $g \rightarrow bb$ candidate, the Xbb2020 tagger performance depends on up to three. Fig. 3.4 shows the variation in D_{Xbb} distributions within each category (with missing jets labelled with an 'x'). While the statistical uncertainty of some of the rarer processes is quite large, the differences within each category are generally smaller than the differences between categories. These five category labels provide sets of events which each have distinct D_{Xbb} distributions, comprise a significant fraction of the total simulated data, and are distinguishable in the fit procedure described below.

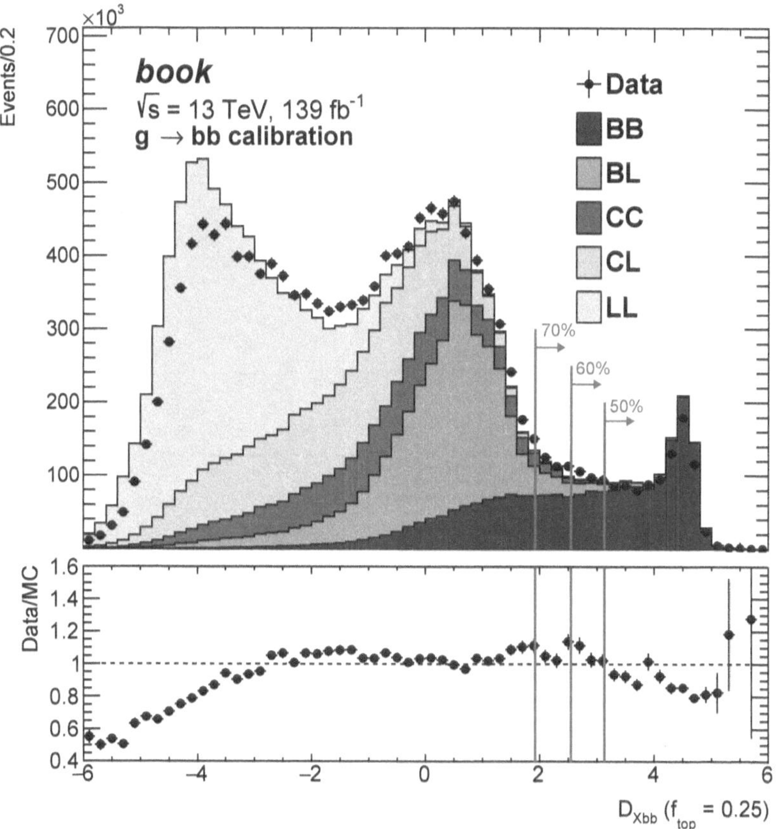

Figure 3.3: The D_{Xbb} distribution, with a top quark fraction of $f_{top} = 0.25$, summed over all p_T bins. Vertical lines show the cuts corresponding to 50%, 60% and 70% efficiency working points.

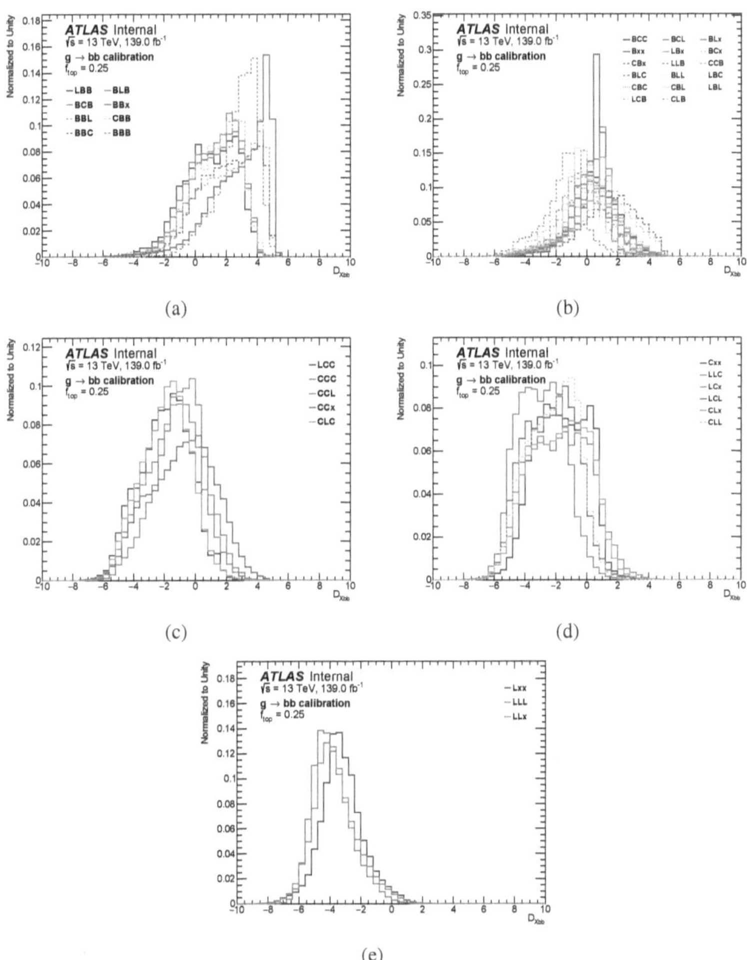

Figure 3.4: The D_{Xbb} distribution of the various sub-categories within each of the (a) *BB*, (b) *BL*, (c) *CC*, (d) *CL* and (e) *LL* flavor categories. These sub-categories include the full information on the flavors of the three track-jets in the $g \to bb$ candidate.

3.4 Flavor Fraction Fit

3.4.1 Template Definition

A template fit method is used to measure the fraction of $g \rightarrow bb$ decays in data. First, a set of templates are constructed using a flavor-sensitive variable. As the template shape differs for each of the flavor categories previously described, fitting the set of templates to the data can extract the relative contributions of each. The variable used in this calibration is the mean of the signed d_0 significance, $\langle s_{d0} \rangle$. The s_{d0} of a track is defined as:

$$s_{d0} = \frac{d_0}{\sigma(d_0)} s_j, \tag{3.3}$$

where $\sigma(d_0)$ is the uncertainty on the d_0 measurement and s_j is the sign of d_0 with respect to the jet axis, i.e. whether the track crosses the jet axis in front of or behind the primary vertex. For jets containing no b-hadron decays, track s_{d0} values are expected to be randomly distributed about zero with a width based on the track angluar resolution. Tracks from secondary vertices have large positive s_{d0} values, and the overall distribution of tracks in b-jets has a large positive tail. For a given jet, the $\langle s_{d0} \rangle$ is defined as the mean of the s_{d0} values of the three highest p_T tracks associated with the jet. This helps reduce the influence of outliers in light-flavor jets from e.g. mis-modelled tracks or K_s decays. Fig. 3.5 shows the $\langle s_{d0} \rangle$ distributions for the muon-jet and leading non-muon-jet within each flavor category. These distributions are binned so that the relative statistical uncertainty is less than 75% in each bin for all templates.

Impact parameter resolution is difficult to accurately model as it depends critically on the resolution of individual hits in the inner detector. This resolution is determined empirically in data and measured in bins of p_T and η using an iterative Gaussian fit procedure described in Ref. [49]. The simulation is then corrected to match the measured impact parameter resolution using a Gaussian smearing function. An additional correction is applied to the simulation to account for a warping of the innermost layer of pixel sensors. The true shape of the IBL was measured using track-to-hit

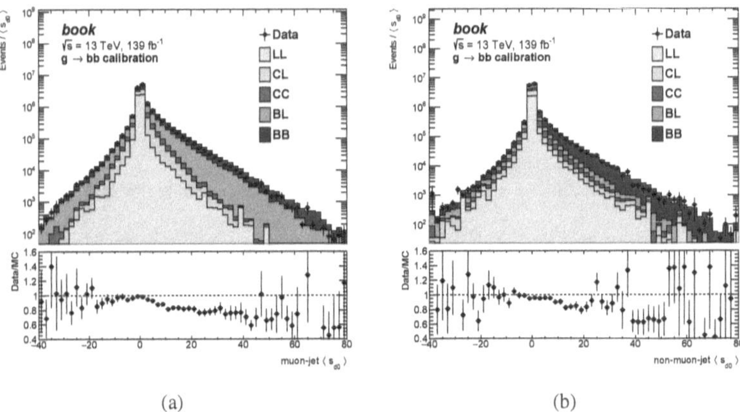

Figure 3.5: The inclusive $\langle s_{d0} \rangle$ distributions for (a) the muon-jet and (b) the non-muon-jet. These distributions include both tagged and anti-tagged events from all p_T bins. Before the fit procedure is applied, some mismodelling is observed in the tails of the distributions.

residuals from a sample of $\sim 2 \times 10^9$ tracks recorded in late 2017 [93]. The correction is only applied to simulations of the 2017 data-taking corrections, as the IBL was inserted during the 2016 end-of-year shutdown. The measured IBL shape was used for simulations of 2018 data-taking conditions, so these simulations require no further correction.

3.4.2   Fit Method

Four sets of templates are used to extract the scale factors: the muon-jet and non-muon-jet $\langle s_{d0} \rangle$ distributions in both the tagged and anti-tagged regions. The fit is done using a binned profile likelihood method. The expectation value in each histogram bin is given by the sum over the flavor templates:

$$E[n_i] = \prod_{xx} f_{xx} y_{xx,i} = f \cdot y_i, \tag{3.4}$$

where $y_{xx,i}$ is the nominal number of entries in bin i from template xx, and f_{xx} is a correction factor to template xx. As in the $HH \rightarrow 4b$ analysis, the template distributions depend on nuisance parameters with prior probability distributions determined from auxiliary measurements. The likelihood function for a histogram can then be constructed as the product of Poisson distributions for each bin and split-normal distributions (denoted by $\mathcal{SN}$) for the NPs:

$$\mathcal{L}(f, \theta) = \prod_{i=1}^{N} e^{-(f \cdot y_i)} \frac{(f \cdot y_i)^{n_i}}{n_i!} \prod_k \mathcal{SN}(\theta_k). \tag{3.5}$$

Maximizing this likelihood function allows one to extract the best-fit values of f or, in other words, measure the fraction of each flavor in the data. The histograms in the tagged and anti-tagged regions are fit simultaneously to extract the scale factor. For statistically independent regions, the combined likelihood is the product of the individual likelihoods:

$$\mathcal{L}(\mu, f(\mu), \theta(\mu)) = \prod_{i=1}^{N} e^{-(E[n_i])} \frac{(E[n_i])^{n_i}}{n_i!} \prod_{j=1}^{M} e^{-(E[m_j])} \frac{(E[m_j])^{m_j}}{m_j!} \prod_k \mathcal{N}(\theta_k) \tag{3.6}$$

where the index i runs over the bins of the tagged region histogram and j runs over the anti-tagged region. The scale factor, $\mu = \epsilon_{\text{data}}/\epsilon_{\text{MC}}$, multiplies the BB template in the tagged region while a corresponding anti-tag scale factor appears in the anti-tagged region. The anti-tag scale factor represents the difference in tagging rejection rates, and is correlated to the scale factor as follows:

$$\mu_{\text{anti-tag}} = \frac{1 - \epsilon_{\text{data}}}{1 - \epsilon_{\text{MC}}} = \frac{1 - \epsilon_{\text{MC}}\mu}{1 - \epsilon_{\text{MC}}}. \tag{3.7}$$

The correlated form of the anti-tag scale factor is used in the likelihood, as shown in Eq. 3.8.

$$\begin{aligned}
E[n_i] &= \mu f_{BB} y_{BB,i} + f_{non-BB} \cdot y_{non-BB,i} \\
E[m_j] &= \frac{1 - \epsilon_{\text{MC}}\mu}{1 - \epsilon_{\text{MC}}} f_{BB} y_{BB,j} + f_{non-BB} \cdot y_{non-BB,j}
\end{aligned} \tag{3.8}$$

In order to allow the fit to distinguish between single-b and double-b decays, the muon-jet and non-muon-jet distributions are fit simultaneuosly. As the shape of the $\langle s_{d0} \rangle$ templates depends only on the flavor of that particular jet, these distributions are uncorrelated and the combined likelihood is simply the product of the individual likelihoods. Nuisance parameters are correlated between the jet distributions and between tagging regions. Each p_{T} bin, however, is fit independently from the rest and the scale factor and flavor corrections are allowed to float freely.

3.5 Systematic Uncertainties

There are many sources of uncertainty considered on the scale factor measurement. These can be grouped into a few categories: experimental uncertainties on the physics objects used, theoretical uncertainties on simulation modelling, uncertainties on the method used to derive the scale factors. Experimental uncertainties are defined by dedicated measurements comparing data to MC simulation and implemented as variations of simulation. The jets, muons and tracks used in the $g \rightarrow bb$ calibration each have a set of associated uncertainties. The largest uncertainty on jet reconstruction comes from the jet energy scale (JES), which quantifies the average difference between the 'true' energy of a jet and the calibrated measurement of the calorimeter. For tracks, on the other hand, the largest uncertainty comes from differences between the efficiency of the track reconstruction algorithm in data and in MC. Each uncertainty adds one or more NPs to the profile-likelihood fit, as described in the previous section. Theoretical uncertainties on the modelling come from the approximations used in the MC simulation. Many of these uncertainties are evaluated using an ad-hoc method in which two datasets, with different approximations, are assumed to fully characterize the space of possible theories. The 'two-point' uncertainties are used where fully sampling the theory space is either computationally or theoretically infeasible. The uncertainties on the parton showering, for example, are evaluated by comparing the results of the PYTHIA 8 and HERWIG 7 generators. Other uncertainties, for example on the renormalization scale of the theory, are evaluated by varying input parameters within a single generator. The theoretical uncertainties are well-defined but are not included in the results presented in this book. The definitions of the theoretical uncertainties and the experimental uncertainties on the physics objects are the same for most ATLAS analyses.

In order to separate effects of the systematics from effects of limited statistical precision, a smoothing function is applied to the systematic impact. Smoothing is applied to the ratio of the nominal and systematic histograms by first merging bins with large statistical uncertainty and then

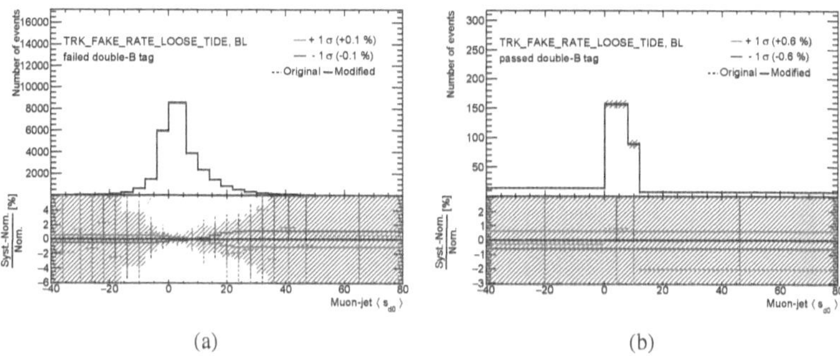

(a)   (b)

Figure 3.6: Template variations due to uncertainty in track reconstruction effiency in the (a) failed tag and (b) pass regions in the [500, 550) GeV p_T bin for the 50% efficiency Xbb2020 working points, with $f_{top} = 0.25$. The solid lines show the variations after smoothing, while dashed points shows the pre-smoothing uncertainties. The statistical uncertainty on the nominal template is shown by the hatched region.

averaging each bin with the average of its neighbors. This type of smoothing fails on the steeply falling template distributions, but it works well on ratios between templates. In particular, this type of smoothing mitigates the effect of individual events with large weights in the systematics-varied templates. The effect of the smoothing is illustrated in Figure 3.6, which shows template variations due to uncertainty in the track reconstruction efficiency. In addition to smoothing, the total number of simulated events predicted by each systematic is set to be the same as that of the nominal prediction. The effects of the systematic on the relative differences between flavors are therefore decoupled from the overall normalization differences between simulation and data.

There are also uncertainties specific to the $g \rightarrow bb$ calibration, primarily related to the template fit method. These include uncertainties on the rates of other processes that produce large s_{d0} tracks, collectively called 's_{d0}' uncertainties, as well as uncertainties on the relative production rates of b-hadrons with different lifetimes. The events used to calculate scale factors differ from events on which they will be applied. Firstly, these events contain $g \rightarrow bb$ decays where searches for new physics will mostly apply the $X \rightarrow bb$ tagging algorithm to $H \rightarrow bb$ decays. Secondly, these events are required to contain muons, a requirement that increases the b-fraction of the sample but also biases the s_{d0} distribution. Extrapolation uncertainties for both are currently being developed, but are not included in the results presented in this book.

3.5.1 s_{d0} Uncertainties

Tracks with large impact parameters can come from a number of sources other than b-hadron decays. These included long-lived species of light hadrons, such as K_s and Λ, photons that convert to e^+e^- pairs in the tracking detector, and particles that change direction after interacting with the detector material. Each of these can potentially create a large s_{d0} track in a light-flavored jet, but the rates of these processes are difficult to measure. In order to estimate the impact of these processes on the calibration, the rates of each are artificially varied up/down by 10% in the simulation. The difference between BB templates with these variations and the nominal templates are smaller than the statistical uncertainties on the templates themselves, as shown in Figure 3.7.

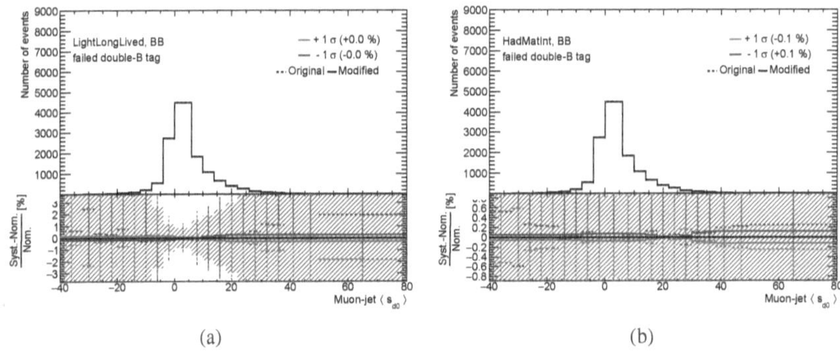

Figure 3.7: Template variations due to increased rates of (a) long-lived light hadrons and (b) hard material interactions in the [500, 550] GeV p_T bin for the 50% efficiency Xbb2020 working points, with $f_{top} = 0.25$. The variations, given by red and blue lines, are smaller than the hatched region showing the statistical uncertainty on the nominal template.

3.5.2 Fake Muons

Muons are produced at higher rates through semi-leptonic b-hadron decays in true $g \rightarrow bb$ events than in the decays of light jets. All events used in the calibration are required to contain a reconstructed muon. Events in the light jet templates mostly contain fakes, however, which can be constructed from the ID track of another object connected to unrelated signals in the muon systems. In the simulation, fake muons are easily identified by the lack of a corresponding muon in the truth record, but the rate of fakes in data is harder to measure. Fully 95% of events in the light jet template do not contain a real muon, while each of the heavy flavor templates contains at most 5% fake muons. As the fake rate has only a small impact on the final result, a conservative uncertainty is estimated by varying the fake rate up/down by 30% in the simulation.

3.5.3 b-hadron Branching Fractions

Differences in lifetimes between different b-hadrons can affect the $\langle s_{d0} \rangle$ template used in the fit. The b-hadron branching fraction uncertainty is intended to account for effects due to mis-modelling of the relative proportions of the b-hadrons in the simulated sample. The Heavy Flavor Averaging Group (HFLAV) publishes calculations of the b-hadronization fractions based on measurements in Z decays from e^+e^- colliders, as well as measurements from pp collisions [94]. Both ATLAS and LHCb have measured the B_s^0/B^0 ratio [95, 96] in $\sqrt{s} = 7\,\mathrm{TeV}$ pp collisions, and found values that agree with the combined Z decay calculation. The full table of fractions from Z decays is therefore used to define this uncertainty, and Table 3.3 shows a comparison to the fractions found in the $g \to b\bar{b}$ sample. An uncertainty is defined by reweighting the b-hadron fractions to match the HFLAV values to create a $+1\sigma$ variation. The difference between the simulated hadronization fractions and the HFLAV values are small, as are the differences in lifetime between the b-hadrons. This uncertainty is therefore expected to be small. It is not, however, included in the results in this book.

b-hadron	in Z decays	in $g \to b\bar{b}$ sample
$B^\pm$	$(40.8 \pm 0.7)\,\%$	$42.2\,\%$
B^0	$(40.8 \pm 0.7)\,\%$	$45.4\,\%$
B_s^0	$(10.0 \pm 0.8)\,\%$	$8.8\,\%$
b-baryon	$(\,8.4 \pm 1.1)\,\%$	$3.7\,\%$

Table 3.3: Production fractions of b-hadrons as calculated from Z decays by HFLAV [94] and in the simulated $g \to b\bar{b}$ decay sample used for the calibration.

3.5.4 Extrapolation Uncertainties

In addition to the uncertainties on the fit itself, uncertainties will also be applied to the scale factor to account for differences between how it was calculated and how it will be used. The Xbb2020 scale factors are intended to be used in analyses searching for resonances decaying to pairs of b quarks. These resonances could be either SM or BSM particles, but will not be the $g \to bb$ process directly. There are some differences between the $g \to bb$ and $H \to bb$ processes, notably in the opening angle between the b-jets and additional gluon radiation in $g \to bb$ decays, and likely similar differences will appear between $g \to bb$ and arbitrary $X \to bb$ signals. The modelling of $g \to bb$ decays in simulation therefore differs from the modelling of $H \to bb$ and the scale factors that account for data-MC differences may not exactly translate from one process to the other. With that said, much of the data-MC difference is expected to come from sources independent of the underlying process and scale factors derived in $g \to bb$ should be applicable to $H \to bb$ decays. Comparisons of the template shapes in simulated $H \to bb$ events will be used to derive an extrapolation uncertainty, although the exact method is not yet defined. This uncertainty is not included in the results in this book.

An extrapolation uncertainty is also needed due to the muon requirement imposed in the $g \to bb$ calibration. Most analyses that wish to use the Xbb2020 tagger will not have an identical requirement, and the presence of a muon can bias the $\langle s_{d0} \rangle$ template. Simulated multijet events that fail the muon requirement are used to estimate this uncertainty by comparing the difference in template shapes between events containing a muon and events that do not. This uncertainty is not yet finalized and is not included in the results in this book.

3.5.5 Summary of Systematics

There are many sources of uncertainty which are, or will be, considered in the $g \to bb$ calibration, and each can affect the result in different ways. Table 3.4 contains a full list of uncertainties, along with brief descriptions. Those uncertainties specifically measured for the analysis have already been described in more depth.

Systematic Uncertainty	Brief Description
Template Uncertainties	
'Fake' secondary vertices	Uncertainty in the rate of processes which create large s_{d0} tracks in light-flavor jets.
'Fake' muons	Uncertainty in the rate of false-positive muon identification in b-jets.
b-hadron fractions	Uncertainty in the relative production rate of b-hadron species. Not included here.
Muon Requirement	Uncertainty in the inclusive phase-space from the muon requirement used to derive scale factors. Not included here.
$g \rightarrow bb$ to $X \rightarrow bb$	Uncertainty in extrapolating from $g \rightarrow bb$ decays to the general $X \rightarrow bb$ case.
Experimental Uncertainties	
Pileup Reweighting	Uncertainties in pile-up conditions are applied when reweighting simulations to match data.
Jet Reconstruction	Uncertainty in the scale and resolution of reconstructed large-R jet energy and mass from detector inputs [87, 88]. Applied as 30 independent NPs on the energy scale, 6 NPs on mass scale and one on Higgs mass resolution.
Muon Reconstruction Efficiency	Uncertainties in the muon reconstruction efficiency and track-to-vertex association [97].
Muon Momentum Scale	Uncertainty in muon momentum reconstruction [97]. Includes separate uncertainties on the resolution of ID and MS tracks.
Sagitta Bias Correction	Uncertainties due to charge-dependent effects of detector mis-alignment [97].
Track reconstruction efficiency	Uncertainties in passive material in the ID and on the GEANT 4 model used in simulation.
Track fake rate	Uncertainty in the rate of combinatorial fake tracks from large numbers of hits in the ID.
Track impact parameter resolution	Uncertainties based on the difference in d_0 and z_0 resolution between data and MC.
Theoretical Uncertainties	
Parton Shower	Uncertainty in the parton shower model is measured by comparing PYTHIA 8 and HERWIG 7. Not included here.
Renormalization Scale	Uncertainties in renormalization and factorization scales, and in final state radiation (FSR) are assessed by sample weight variations in PYTHIA 8. Not included here.

Table 3.4: Uncertainties applied in the derivation of b-tagging scale factors, though some are not yet included in the result.

3.6 Results

The primary result of the $g \rightarrow bb$ calibration is a set of scale factors, and associated uncertainties, that can be applied to adjust the MC simulation to match real data. Preliminary versions of these scale factors are presented below, as a function of large-R jet p_T, for several $X \rightarrow bb$ tagger working points. Figure 3.8 shows the scale factors for the 50%, 60% and 70% efficiency working points, respectively. For each, the $X \rightarrow bb$ tagger discriminant is calculated with f_{top} = 0.25. A general trend is observed for the 50% and 60% working points where the simulation underestimates the tagging efficiency of the algorithm at low p_T and overestimates the efficiency at high p_T. At the 70% efficiency working point, the derived scale factors are all compatible with one. There are currently a few bins for each working point where the maximum-likelihood fit does not converge, always in the bins with $p_T < 500\,\mathrm{GeV}$.

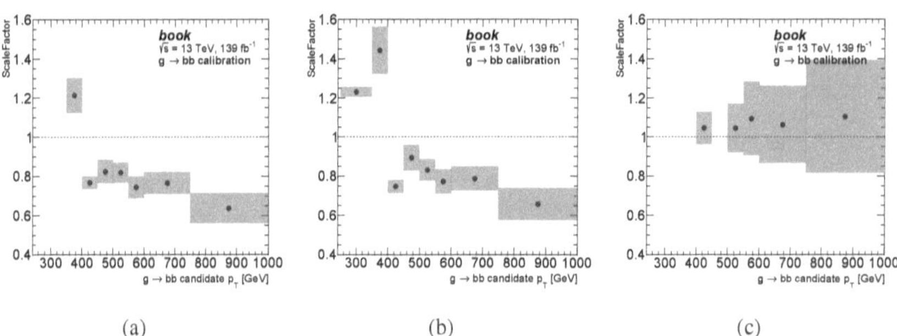

(a)   (b)   (c)

Figure 3.8: Derived scale factors for the (a) 50%, (b) 60%, and (c) 70% efficiency Xbb2020 working points, with f_{top} = 0.25. Results are preliminary and not all uncertainties are currently accounted for. Bins where the maximum-likelihood fit does not converge are left empty in the plot.

The stability and validity of the fit is assessed in multiple ways. The first such check is that after the fit the simulation and data should agree within uncertainties. Figure 3.12 shows the post-fit $\langle s_{d0} \rangle$ templates in the [500, 550) GeV p_T bin for the 50% Xbb2020 WP. After the fit, good agreement is indeed observed between data and the normalized flavor templates. Some discrepancy remains, however, in events that pass the tagging, where the BB template contains more high-$\langle s_{d0} \rangle$ events than are seen in data. Preliminary studies on samples simulated with HERWIG 7 indicate

111

that this mis-modelling may come from the parton shower model. These studies are still in an early
stage and are not presented in this book.

Another check of the fit consistents of looking at the pulls and impacts of the NPs [1]. As in
the $HH \rightarrow 4b$ analysis, pulls are generally expected to be less than 1σ from the nominal value of
zero, with the exception of the template normalizations, $f(\mu)$. Figure 3.9 shows the flavor template
normalizations for each working point. Flavor normalizations as large as 100% are observed in
some p_T bins, indicating significant mismodeling of the flavor fractions in the multijet simulation.
Large (anti-)correlations between the flavor corrections and the scale factors are expected, and
observed. Figure 3.10 shows these correlations for the [500, 550) GeV p_T bin of the 50% WP.

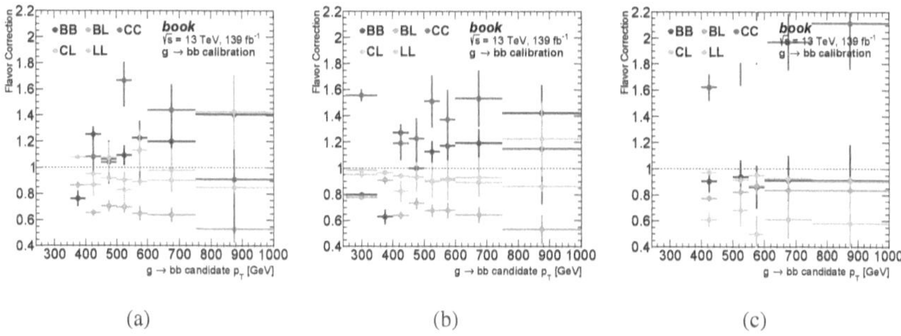

(a)   (b)   (c)

Figure 3.9: Derived flavor correction factors for the (a) 50%, (b) 60%, and (c) 70% efficiency
Xbb2020 working points, with $f_{\text{top}} = 0.25$.

Figure 3.11 compares the prior and posterior distributions for the NPs in each fit, in the [500,
550) GeV p_T bin. The best-fit value for most NPs is within 1σ of the initial value in most cases
but some NPs deviate significantly from zero. Most notably the track reconstruction efficiency, as
well as the rate of long-lived light hadrons, show significant pulls. these NPs have the effect of
decreasing the number of high-$\langle s_{d0} \rangle$ events in the BL and BB templates respectively. The observed
pulls may change as the final uncertainties are added to the fit.

---

[1] Recall that the pull is the difference in mean between the prior and posterior NP distributions, and the impact is
the effect a 1σ variation of the NP has on the measured scale factor.

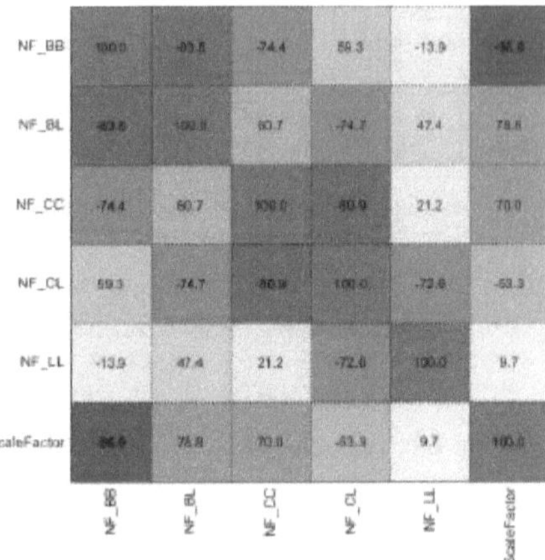

Figure 3.10: Correlations between the flavor corrections and the scale factor in the [500, 550) GeV p_T bin for the 50% Xbb2020 WP, with $f_{\text{top}} = 0.25$.

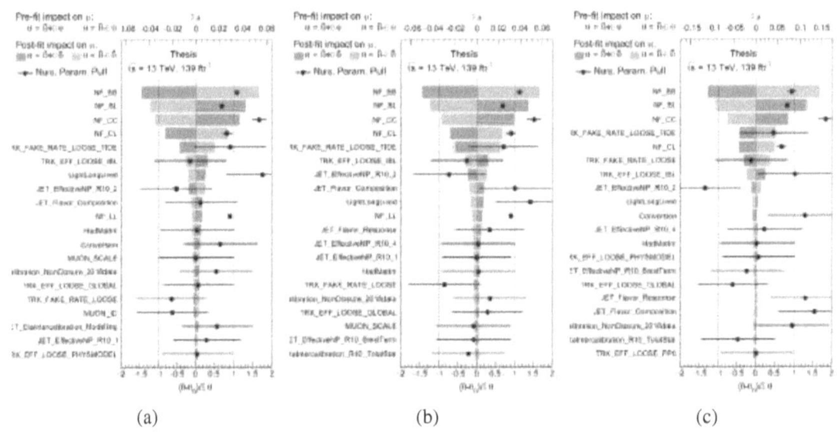

Figure 3.11: Pulls (points, bottom axis) and impacts (bars, top axis) of nuisance parameters included in the fits in the [500, 550) GeV p_T bins for the (a) 50%, (b) 60%, and (c) 70% efficiency Xbb2020 working points, with $f_{\text{top}} = 0.25$. The flavor corrections are distributed about one, while other NPs are distributed about zero.

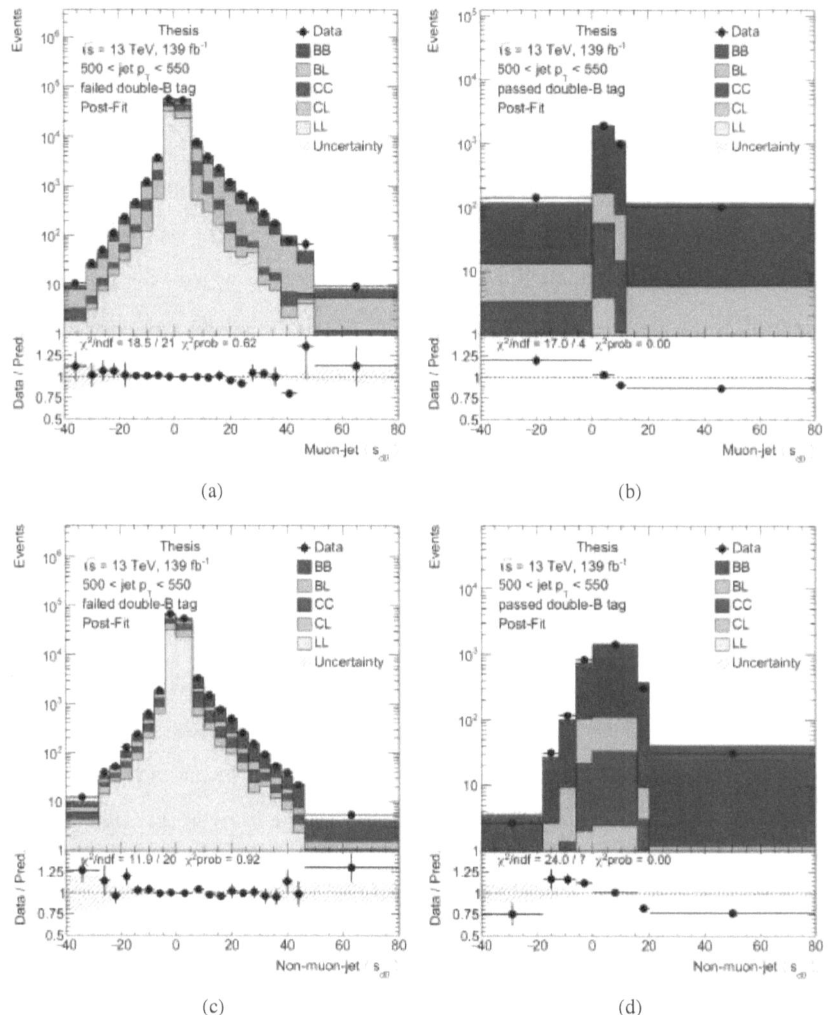

Figure 3.12: Post-fit $\langle s_{d0} \rangle$ template distributions in the [500, 550) GeV p_T bin for the 50% Xbb2020 WP, with $f_{top} = 0.25$. Events that fail the b-tagging cut are shown in (a) and (c) while those that pass are shown in (b) and (d). Muon-jet distributions are shown in (a) and (b), and non-muon-jet distributions are shown in (c) and (d).

114

3.7 Next Steps

The $g \to b\bar{b}$ calibration is not yet ready for use in ATLAS analyses, but preliminary results are promising. A few open questions remain about the stability of the template fit, and the extrapolation from the phase space of the calibration to the case of an arbitrary analysis. In addition, a few sources of systematic uncertainty are not yet accounted for in the fit. These concerns are being addressed by other students, and we hope to have a complete result ready soon. The final results of the calibration will be sets of scale factors similar to those presented in Figure 3.8. In parallel, calibrations of the Xbb2020 tagger efficiency using $Z \to b\bar{b}$ decays are being studied in both Z+jets and $Z+\gamma$ final states, and a calibration of the hadronic top quark mis-tag rate is being studied in semileptonic top quark decays. Ultimately these separate efforts may be combined to provide more accurate scale factors than any of the calibrations individually. Once the calibrations are ready, the Xbb2020 algorithm will be ready to use in searches for new physics, bringing improvements to a number of analyses including, potentially, future searches for resonant $HH \to 4b$ decays.

3.7.1 Using Xbb2020 in $HH \to 4b$

While the Xbb2020 algorithm was not used in the latest $HH \to 4b$ analysis, it, or an improved version of it, may be used in the next. Simple modifications of the current analysis strategy could be made to update the tagging strategy to use double-b-taggers. I present here an example for comparison purposes. The current strategy of separating events into three b-tagging channels outperforms any individual Xbb2020 cut, as the limits are defined primarily by the best channel for each mass point. In order to estimate the potential improvement from using Xbb2020, I therefore define three exclusive Xbb2020 channels. The Xbb2020 cuts are chosen to have similar background rejection to the cuts used to define $4b$, $3b$, and $2b$-split channels in order to compare the signal efficiency of the algorithms directly. The set of cuts used are listed in Table 3.5, with background rejection measured using simulated multijet and $t\bar{t}$ events. The signal efficiency, as a function of mass, is shown in Figure 3.13.

DL1r channel	Xbb2020 cut ($f_{\text{top}} = 0.25$)	Background rejection
$4b$	$D_{\text{Xbb}} > 3.15$	14800
$3b$	$0.93 < D_{\text{Xbb}} < 3.15$	670
$2b$-split	$-0.38 < D_{\text{Xbb}} < 0.93$	110

Table 3.5: A set of exclusive Xbb2020 cuts, with $f_{\text{top}} = 0.25$, that reject similar proportions of simulated multijet and $t\bar{t}$ backgrounds to the $4b$, $3b$, and $2b$-split channels defined using DL1r.

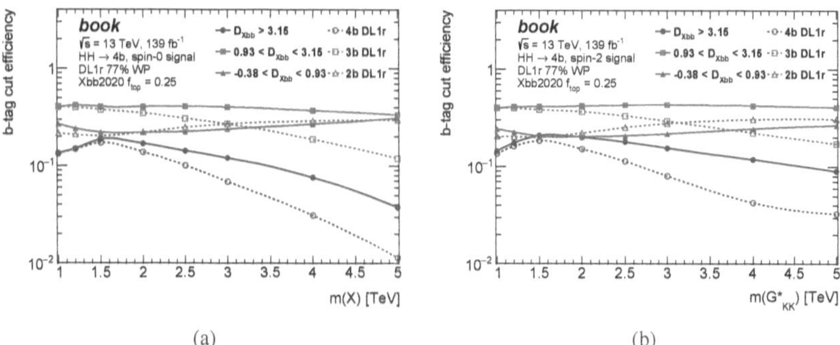

(a)   (b)

Figure 3.13: Efficiency of Xbb2020 cuts applied to the (a) spin-0 and (b) spin-2 $HH \rightarrow 4b$ signal models. The set of Xbb2020 cuts where chosen to have background rejection comparable to the $4b$, $3b$, and $2b$-split channels presented in Section 2.3.

The Xbb2020 algorithm provides improved signal efficiency for high mass signals with the same background rejection, outperforming the current strategy that uses the DL1r algorithm. While simulated events were used as for a simplified background estimate in this study, one could imagine a data-driven background model for the Xbb2020 channels defined using 'low-tag' regions in which one of the Higgs candidate jets fails the b-tagging. Of the current steps in the background modelling process, the kinematic reweighting would likely change the most when changing to a new b-tagging paradigm. The cuts chosen here were simply for comparison and further improvement may be gained by a dedicated optimization of Xbb2020 regions. Even this simple modifications shows the potential gains in signal efficiency at high mass made possible by the Xbb2020 algorithm.

116

Chapter 4: Conclusion

Two results have been presented in this book: the search for heavy resonances decaying to $HH \rightarrow 4b$ in 139 fb^{-1} of ATLAS data, and the $g \rightarrow bb$ calibration of the $X \rightarrow bb$ double-b-tagging algorithm. The $HH \rightarrow 4b$ analysis tested two signal models, searching for evidence of a spin-0 or spin-2 resonance with a mass of 251-5000 GeV. No significant excesses above the Standard Model prediction were observed, therefore upper limits were set on the production cross-section of spin-0 and spin-2 resonances. In addition the RS model was excluded for gravitons with mass between 298 GeV and 1440 GeV. Searches for resonant HH production in other final states using the full Run 2 dataset are being conducted, with the boosted $b\bar{b}\tau^+\tau^-$ analysis already published. A combination of these results is planned and is expected to significantly improve the limits in the resolved regime. $HH \rightarrow 4b$ remains a promising channel to search for new physics, as many BSM theories alter the properties of the Higgs boson. The LHC, as well as the ATLAS detector, shut down for upgrades at the end 2018, marking the end of Run 2. The start of Run 3 has been delayed by the COVID-19 pandemic, and Run 3 is now planned to last from May 2022 to October 2024, doubling the ATLAS proton–proton collision dataset. In total, ATLAS is expected to collected around 3 ab^{-1} over the lifetime of the LHC. Searches for high mass resonances, and searches with large multijet backgrounds, are generally limited primarily by a lack of data, and this is certainly the case for the $HH \rightarrow 4b$ analysis. The $HH \rightarrow 4b$ analysis will likely be redone periodically as new data is collected, to search for new evidence or improve the limits set here. While the calibration of the Xbb2020 algorithm was not ready in time for this $HH \rightarrow 4b$ search, future versions of the analysis will likely use dedicated double-b-taggers.

For the moment, no conclusive evidence of BSM physics has been discovered by the LHC experiments. The search presented in this book found no such evidence but set new limits on phase space of these theories. New techniques and a larger dataset allowed this search to significantly improve the limits set by previous analyses. Calibration development for double-b-tagging paves the way for future refinement of these techniques, and further improvement can be expected as ATLAS collects more data. Limits set by the ATLAS collaboration, including those presented here, continue to constrain and guide the search for physics beyond the Standard Model.

References

[1]   ATLAS Collaboration, "Observation of a new particle in the search for the Standard Model Higgs boson with the ATLAS detector at the LHC," Phys. Lett. B, vol. 716, p. 1, 2012. arXiv: 1207.7214 [hep-ex].

[2]   CMS Collaboration, "Observation of a new boson at a mass of 125 GeV with the CMS experiment at the LHC," Phys. Lett. B, vol. 716, p. 30, 2012. arXiv: 1207.7235 [hep-ex].

[3]   Particle Data Group, "Review of Particle Physics," PTEP, vol. 2020, no. 8, p. 083C01, 2020.

[4]   J. Ellis, "Higgs Physics," 117–168. 52 p, 2013, 52 pages, 45 figures, Lectures presented at the ESHEP 2013 School of High-Energy Physics, to appear as part of the proceedings in a CERN Yellow Report. arXiv: 1312.5672.

[5]   L. Randall and R. Sundrum, "A Large mass hierarchy from a small extra dimension," Phys. Rev. Lett., vol. 83, pp. 3370–3373, 1999. arXiv: hep-ph/9905221.

[6]   K. Agashe, H. Davoudiasl, G. Perez, and A. Soni, "Warped Gravitons at the LHC and Beyond," Phys. Rev. D, vol. 76, p. 036006, 2007. arXiv: hep-ph/0701186.

[7]   A. Fitzpatrick, J. Kaplan, L. Randall, and L.-T. Wang, "Searching for the Kaluza-Klein Graviton in Bulk RS Models," JHEP, vol. 09, p. 013, 2007. arXiv: hep-ph/0701150.

[8]   A. Oliveira, "Gravity particles from Warped Extra Dimensions, predictions for LHC," Mar. 2014. arXiv: 1404.0102 [hep-ph].

[9]   G. Branco, P. Ferreira, L. Lavoura, M. Rebelo, M. Sher, and J. P. Silva, "Theory and phenomenology of two-Higgs-doublet models," Phys. Rept., vol. 516, pp. 1–102, 2012. arXiv: 1106.0034 [hep-ph].

[10] A. Djouadi and J. Quevillon, "The MSSM Higgs sector at a high M_{SUSY}: reopening the low tanβ regime and heavy Higgs searches," JHEP, vol. 10, p. 028, 2013. arXiv: 1304.1787 [hep-ph].

[11] A. Djouadi, L. Maiani, G. Moreau, A. Polosa, J. Quevillon, and V. Riquer, "The post-Higgs MSSM scenario: Habemus MSSM?" Eur. Phys. J. C, vol. 73, p. 2650, 2013. arXiv: 1307.5205 [hep-ph].

[12] L. Evans and P. Bryant, "LHC Machine," JINST, vol. 3, S08001, 2008.

[13] P. Mouche, "Overall view of the LHC. Vue d'ensemble du LHC," 2014, General Photo.

[14] E. Todesco and J. Wenninger, "Large hadron collider momentum calibration and accuracy," Phys. Rev. Accel. Beams, vol. 20, p. 081 003, 2017.

[15] M. Dobbs and J. B. Hansen, "The HepMC C++ Monte Carlo event record for High Energy Physics," Comput. Phys. Commun., vol. 134, pp. 41–46, 2001.

[16] V. N. Gribov and L. N. Lipatov, "Deep inelastic e p scattering in perturbation theory," Sov. J. Nucl. Phys., vol. 15, pp. 438–450, 1972.

[17] G. Altarelli and G. Parisi, "Asymptotic Freedom in Parton Language," Nucl. Phys. B, vol. 126, pp. 298–318, 1977.

[18] J. Pequenao, "Computer generated image of the whole ATLAS detector," 2008.

[19] ATLAS Collaboration, "The ATLAS Experiment at the CERN Large Hadron Collider," JINST, vol. 3, S08003, 2008.

[20] Abbott, B. and others, "Production and integration of the ATLAS Insertable B-Layer," JINST, vol. 13, T05008, 2018. arXiv: 1803.00844 [physics.ins-det].

[21] ATLAS Collaboration, "ATLAS Insertable B-Layer: Technical Design Report," ATLAS-TDR-19; CERN-LHCC-2010-013, 2010.

[22] ATLAS Collaboration, "ATLAS Liquid Argon Calorimeter: Technical Design Report," ATLAS-TDR-2; CERN-LHCC-96-041, 1996.

[23] ATLAS Collaboration, "ATLAS Tile Calorimeter: Technical Design Report," ATLAS-TDR-3; CERN-LHCC-96-042, 1996.

[24] ATLAS Collaboration, "ATLAS Muon Spectrometer: Technical Design Report," CERN, ATLAS-TDR-10; CERN-LHCC-97-022, 1997.

[25] LHC Higgs Cross Section Working Group, S. Dittmaier, C. Mariotti, G. Passarino, and R. Tanaka (Eds.), "Handbook of LHC Higgs Cross Sections: 1. Inclusive Observables," CERN-2011-002, CERN, Geneva, 2011. arXiv: 1101.0593 [hep-ph].

[26] ATLAS Collaboration, "Performance of the ATLAS trigger system in 2015," Eur. Phys. J. C, vol. 77, p. 317, 2017. arXiv: 1611.09661 [hep-ex].

[27] ATLAS Collaboration, "Performance of the ATLAS track reconstruction algorithms in dense environments in LHC Run 2," Eur. Phys. J. C, vol. 77, p. 673, 2017. arXiv: 1704.07983 [hep-ex].

[28] ATLAS Collaboration, "Reconstruction of primary vertices at the ATLAS experiment in Run 1 proton–proton collisions at the LHC," Eur. Phys. J. C, vol. 77, p. 332, 2017. arXiv: 1611.10235 [hep-ex].

[29] ATLAS Collaboration, "Topological cell clustering in the ATLAS calorimeters and its performance in LHC Run 1," Eur. Phys. J. C, vol. 77, p. 490, 2017. arXiv: 1603.02934 [hep-ex].

[30] A. Collaboration, "ATLAS event display: candidate pair of Higgs bosons decay in ATLAS," General Photo, 2021.

[31] J. Pequenao and P. Schaffner, "How ATLAS detects particles: diagram of particle paths in the detector," 2013.

[32] ATLAS Collaboration, "Muon reconstruction and identification efficiency in ATLAS using the full Run 2 pp collision data set at $\sqrt{s}$ = 13 TeV," 2020. arXiv: 2012.00578 [hep-ex].

[33] S. D. Ellis and D. E. Soper, "Successive combination jet algorithm for hadron collisions," Phys. Rev. D, vol. 48, pp. 3160–3166, 1993. arXiv: hep-ph/9305266.

[34] M. Cacciari, G. P. Salam, and G. Soyez, "FastJet user manual," Eur. Phys. J. C, vol. 72, p. 1896, 2012. arXiv: 1111.6097 [hep-ph].

[35] M. Cacciari, G. P. Salam, and G. Soyez, "The anti-k_t jet clustering algorithm," JHEP, vol. 04, p. 063, 2008. arXiv: 0802.1189 [hep-ph].

[36] ATLAS Collaboration, "Identification and rejection of pile-up jets at high pseudorapidity with the ATLAS detector," Eur. Phys. J. C, vol. 77, p. 580, 2017. arXiv: 1705.02211 [hep-ex].

[37] D. Krohn, J. Thaler, and L.-T. Wang, "Jet Trimming," JHEP, vol. 02, p. 084, 2010. arXiv: 0912.1342 [hep-ph].

[38] ATLAS Collaboration, "Optimisation of large-radius jet reconstruction for the ATLAS detector in 13 TeV proton–proton collisions," 2020. arXiv: 2009.04986 [hep-ex].

[39] D. Krohn, J. Thaler, and L.-T. Wang, "Jets with Variable R," JHEP, vol. 06, p. 059, 2009. arXiv: 0903.0392 [hep-ph].

[40] ATLAS Collaboration, "Variable Radius, Exclusive-k_T, and Center-of-Mass Subjet Reconstruction for Higgs($\rightarrow b\bar{b}$) Tagging in ATLAS," 2017.

[41] M. Cacciari and G. P. Salam, "Pileup subtraction using jet areas," Phys. Lett. B, vol. 659, pp. 119–126, 2008. arXiv: 0707.1378 [hep-ph].

[42] ATLAS Collaboration, "Optimisation and performance studies of the ATLAS b-tagging algorithms for the 2017-18 LHC run," 2017.

[43] ATLAS Collaboration, "Identification of Jets Containing b-Hadrons with Recurrent Neural Networks at the ATLAS Experiment," 2017.

[44] ATLAS Collaboration, "Secondary vertex finding for jet flavour identification with the ATLAS detector," 2017.

[45] ATLAS Collaboration, "Topological b-hadron decay reconstruction and identification of b-jets with the JetFitter package in the ATLAS experiment at the LHC," 2018.

[46] ATLAS Collaboration, "ATLAS b-jet identification performance and efficiency measurement with $t\bar{t}$ events in pp collisions at $\sqrt{s}$ = 13 TeV," 2019. arXiv: 1907.05120 [hep-ex].

[47] ATLAS Collaboration, "Optimisation of the ATLAS b-tagging performance for the 2016 LHC Run," 2016.

[48] ATLAS Collaboration, "Performance of 2019 recommendations of ATLAS Flavor Tagging algorithms with Variable Radius track jets," 2019.

[49] ATLAS Collaboration, "Calibration of light-flavour b-jet mistagging rates using ATLAS proton–proton collision data at $\sqrt{s}$ = 13 TeV," 2018.

[50] ATLAS Collaboration, "Measurement of b-tagging efficiency of c-jets in $t\bar{t}$ events using a likelihood approach with the ATLAS detector," 2018.

[51] J. Alwall, R. Frederix, S. Frixione, V. Hirschi, F. Maltoni, O. Mattelaer, H. S. Shao, T. Stelzer, P. Torrielli, and M. Zaro, "The automated computation of tree-level and next-to-leading order differential cross sections, and their matching to parton shower simulations," JHEP, vol. 07, p. 079, 2014. arXiv: 1405.0301 [hep-ph].

[52] J. Bellm et al., "Herwig 7.1 Release Note," 2017. arXiv: 1705.06919 [hep-ph].

[53] D. J. Lange, "The EvtGen particle decay simulation package," Nucl. Instrum. Meth. A, vol. 462, p. 152, 2001.

[54] Harland-Lang, L.A. and Martin, A.D. and Motylinski, P. and Thorne, R.S., "Parton distributions in the LHC era: MMHT 2014 PDFs," p. 83, 2014. arXiv: 1412.3989 [hep-ph].

[55] T. Sjöstrand, S. Ask, J. R. Christiansen, R. Corke, N. Desai, P. Ilten, S. Mrenna, S. Prestel, C. O. Rasmussen, and P. Z. Skands, "An introduction to PYTHIA 8.2," Comput. Phys. Commun., vol. 191, p. 159, 2015. arXiv: 1410.3012 [hep-ph].

[56] ATLAS Collaboration, "ATLAS Pythia 8 tunes to 7 TeV data," 2014.

[57] R. D. Ball et al., "Parton distributions with LHC data," Nucl. Phys. B, vol. 867, p. 244, 2013. arXiv: 1207.1303 [hep-ph].

[58] S. Agostinelli et al., "GEANT4 – a simulation toolkit," Nucl. Instrum. Meth. A, vol. 506, p. 250, 2003.

[59] P. Nason, "A new method for combining NLO QCD with shower Monte Carlo algorithms," JHEP, vol. 11, p. 040, 2004. arXiv: hep-ph/0409146.

[60] S. Frixione, P. Nason, and C. Oleari, "Matching NLO QCD computations with parton shower simulations: the POWHEG method," JHEP, vol. 11, p. 070, 2007. arXiv: 0709.2092 [hep-ph].

[61] S. Alioli, P. Nason, C. Oleari, and E. Re, "A general framework for implementing NLO calculations in shower Monte Carlo programs: the POWHEG BOX," JHEP, vol. 06, p. 043, 2010. arXiv: 1002.2581 [hep-ph].

[62] J. M. Campbell, R. K. Ellis, P. Nason, and E. Re, "Top-Pair Production and Decay at NLO Matched with Parton Showers," JHEP, vol. 04, p. 114, 2015. arXiv: 1412.1828 [hep-ph].

[63] ATLAS Collaboration, "Studies on top-quark Monte Carlo modelling for Top2016," Oct. 2016.

[64] M. Grazzini, G. Heinrich, S. Jones, S. Kallweit, M. Kerner, J. M. Lindert, and J. Mazzitelli, "Higgs boson pair production at NNLO with top quark mass effects," JHEP, vol. 05, p. 059, 2018. arXiv: 1803.02463 [hep-ph].

[65] J. Baglio, F. Campanario, S. Glaus, M. Mühlleitner, J. Ronca, and M. Spira, "$gg \to HH$: Combined uncertainties," Phys. Rev. D, vol. 103, no. 5, p. 056002, 2021. arXiv: 2008.11626 [hep-ph].

[66] CMS Collaboration, "The CMS experiment at the CERN LHC," JINST, vol. 3, S08004, 2008.

[67] ATLAS Collaboration, "Search for the $HH \to b\bar{b}b\bar{b}$ process via vector-boson fusion production using proton–proton collisions at $\sqrt{s} = 13\,\text{TeV}$ with the ATLAS detector," JHEP, vol. 07, p. 108, 2020. arXiv: 2001.05178 [hep-ex].

[68] ATLAS Collaboration, "Search for pair production of Higgs bosons in the $b\bar{b}b\bar{b}$ final state using proton–proton collisions at $\sqrt{s} = 13\,\text{TeV}$ with the ATLAS detector," Phys. Rev. D, vol. 94, p. 052002, 2016. arXiv: 1606.04782 [hep-ex].

[69] ATLAS Collaboration, "Search for pair production of Higgs bosons in the $b\bar{b}b\bar{b}$ final state using proton–proton collisions at $\sqrt{s}$ = 13 TeV with the ATLAS detector," JHEP, vol. 01, p. 030, 2019. arXiv: 1804.06174 [hep-ex].

[70] ATLAS Collaboration, "A search for resonant and non-resonant Higgs boson pair production in the $b\bar{b}\tau^+\tau^-$ decay channel in pp collisions at $\sqrt{s}$ = 13 TeV with the ATLAS detector," Phys. Rev. Lett., vol. 121, p. 191 801, 2018. arXiv: 1808.00336 [hep-ex].

[71] ATLAS Collaboration, "Search for Higgs boson pair production in the $\gamma\gamma b\bar{b}$ final state with 13 TeV pp collision data collected by the ATLAS experiment," JHEP, vol. 11, p. 040, 2018. arXiv: 1807.04873 [hep-ex].

[72] ATLAS Collaboration, "Search for Higgs boson pair production in the $b\bar{b}WW^*$ decay mode at $\sqrt{s}$ = 13 TeV with the ATLAS detector," JHEP, vol. 04, p. 092, 2019. arXiv: 1811.04671 [hep-ex].

[73] ATLAS Collaboration, "Search for Higgs boson pair production in the $\gamma\gamma WW^*$ channel using pp collision data recorded at $\sqrt{s}$ = 13 TeV with the ATLAS detector," Eur. Phys. J. C, vol. 78, p. 1007, 2018. arXiv: 1807.08567 [hep-ex].

[74] ATLAS Collaboration, "Search for Higgs boson pair production in the $WW^{(*)}WW^{(*)}$ decay channel using ATLAS data recorded at $\sqrt{s}$ = 13 TeV," JHEP, vol. 05, p. 124, 2019. arXiv: 1811.11028 [hep-ex].

[75] ATLAS Collaboration, "Combination of searches for Higgs boson pairs in pp collisions at $\sqrt{s}$ = 13 TeV with the ATLAS detector," 2019. arXiv: 1906.02025 [hep-ex].

[76] CMS Collaboration, "Search for resonant pair production of Higgs bosons decaying to bottom quark-antiquark pairs in proton–proton collisions at 13 TeV," JHEP, vol. 08, p. 152, 2018. arXiv: 1806.03548 [hep-ex].

[77] CMS Collaboration, "Search for Higgs boson pair production in the $\gamma\gamma b\bar{b}$ final state in pp collisions at $\sqrt{s}$ = 13 TeV," Phys. Lett. B, vol. 788, p. 7, 2019. arXiv: 1806.00408 [hep-ex].

[78] CMS Collaboration, "Search for Higgs boson pair production in events with two bottom quarks and two tau leptons in proton–proton collisions at $\sqrt{s}$ = 13 TeV," Phys. Lett. B, vol. 778, p. 101, 2018. arXiv: 1707.02909 [hep-ex].

[79] CMS Collaboration, "Search for resonant and nonresonant Higgs boson pair production in the $b\bar{b}\ell\nu\ell\nu$ final state in proton–proton collisions at $\sqrt{s}$ = 13 TeV," JHEP, vol. 01, p. 054, 2018. arXiv: 1708.04188 [hep-ex].

[80] CMS Collaboration, "Search for resonant pair production of Higgs bosons in the $bbZZ$ channel in proton–proton collisions at $\sqrt{s}$ = 13 TeV," Phys. Rev. D, vol. 102, p. 032003, 2020. arXiv: 2006.06391 [hep-ex].

[81] CMS Collaboration, "Combination of Searches for Higgs Boson Pair Production in Proton–Proton Collisions at $\sqrt{s}$ = 13 TeV," Phys. Rev. Lett., vol. 122, p. 121803, 2019. arXiv: 1811.09689 [hep-ex].

[82] ATLAS Collaboration, "Jet reconstruction and performance using particle flow with the ATLAS Detector," Eur. Phys. J. C, vol. 77, p. 466, 2017. arXiv: 1703.10485 [hep-ex].

[83] ATLAS Collaboration, "Jet mass reconstruction with the ATLAS Detector in early Run 2 data," 2016.

[84] ATLAS Collaboration, "Search for low-scale gravity signatures in multi-jet final states with the ATLAS detector at $\sqrt{s}$ = 8 TeV," JHEP, vol. 07, p. 032, 2015. arXiv: 1503.08988 [hep-ex].

[85] G. Avoni et al., "The new LUCID-2 detector for luminosity measurement and monitoring in ATLAS," JINST, vol. 13, no. 07, P07017, 2018.

[86] ATLAS Collaboration, "Luminosity determination in pp collisions at $\sqrt{s}$ = 13 TeV using the ATLAS detector at the LHC," 2019.

[87] ATLAS Collaboration, "In situ calibration of large-radius jet energy and mass in 13 TeV proton–proton collisions with the ATLAS detector," Eur. Phys. J. C, vol. 79, p. 135, 2019. arXiv: 1807.09477 [hep-ex].

[88] ATLAS Collaboration, "Measurement of the ATLAS Detector Jet Mass Response using Forward Folding with $80\,\mathrm{fb}^{-1}$ of $\sqrt{s}$ = 13 TeV pp data," 2020.

[89] G. Cowan, K. Cranmer, E. Gross, and O. Vitells, "Asymptotic formulae for likelihood-based tests of new physics," Eur. Phys. J. C, vol. 71, p. 1554, 2011. arXiv: 1007.1727 [physics.data-an].

[90] ATLAS Collaboration, "Identification of Boosted Higgs Bosons Decaying Into $b\bar{b}$ With Neural Networks and Variable Radius Subjets in ATLAS," Jul. 2020.

[91] ATLAS Collaboration, "Performance of b-jet identification in the ATLAS experiment," JINST, vol. 11, P04008, 2016. arXiv: 1512.01094 [hep-ex].

[92] ATLAS Collaboration, "Early Inner Detector Tracking Performance in the 2015 data at $\sqrt{s}$ = 13 TeV," 2015.

[93] ATLAS Collaboration, "Alignment of the ATLAS Inner Detector in Run-2," Eur. Phys. J. C, vol. 80, p. 1194, 2020. arXiv: 2007.07624 [hep-ex].

[94] Amhis, Yasmine Sara and others, "Averages of b-hadron, c-hadron, and τ-lepton properties as of 2018," Eur. Phys. J. C, vol. 81, no. 3, p. 226, 2021, updated results and plots available at https://hflav.web.cern.ch/. arXiv: 1909.12524 [hep-ex].

[95] ATLAS Collaboration, "Determination of the ratio of b-quark fragmentation fractions f_s/f_d in pp collisions at $\sqrt{s}$ = 7 TeV with the ATLAS detector," Phys. Rev. Lett., vol. 115, no. 26, p. 262 001, 2015. arXiv: 1507.08925 [hep-ex].

[96] LHCb Collaboration, "Measurement of the fragmentation fraction ratio f_s/f_d and its dependence on B meson kinematics," JHEP, vol. 04, p. 001, 2013. arXiv: 1301.5286 [hep-ex].

[97] ATLAS Collaboration, "Muon reconstruction performance of the ATLAS detector in proton–proton collision data at $\sqrt{s}$ = 13 TeV," Eur. Phys. J. C, vol. 76, p. 292, 2016. arXiv: 1603.05598 [hep-ex].

Appendix : Appendices

A $HH \rightarrow 4b$ Cut Optimization

The event selection cuts used in the $HH \rightarrow 4b$ analysis were optimized using several different methods, depending on the cut. The cuts that were optimized were: the $|\Delta\eta|$ cut, the signal region X_{HH} cut, and the b-tagging cut. While the event selection also includes cuts on large-R jet p_T, mass and η, these cuts are set to based on the trigger and detector geometry rather than rejected background events. Two optimization methods were attempted, which differ in ease of computation and expected validity of the results.

The first optimization method was based on the significance estimate $Z = \sqrt{2((s + b)\ln(1 + s/b) - s)}$, where s is the number of signal events and b is the number of background events. This estimate approximates the median significance of the nominal ($\mu = 1$) signal hypothesis from a counting experiment. The result is expected to differ from the statistical procedure described in Section 2.6 because the shape of the signal and background distributions are lost, as are the nuisance parameters corresponding to systematic uncertainties. In an attempt to correct for the loss of shape information, the total signal and background events were calculated separately for each signal, integrating over a window of $m_{HH} \in (0.75\ m_X, 1.15\ m_X)$, where m_X is the true signal mass. This window removes the low-m_{HH} background events that would otherwise overwhelm the significance estimate. A simplified background estimate was used for this optimization, consisting of simulated $t\bar{t}$ and multijet events, in order to look at the signal regions while keeping data blinded.

The first method was used to optimize the $|\Delta\eta|$ and X_{HH} cuts. For the $|\Delta\eta|$ optimization, a grid of cut values was tested for a set of signal masses and the optimal cuts as a function of mass are shown in Figure 1. Similar results are observed for the spin-0 and spin-2 signal hypotheses. The result of a similar grid scan of X_{HH} cuts is shown in Figure 3a. In addition, a set of alternate signal region shapes and cuts were tested. The X_{HH} function used, Eq. 2.2, is designed

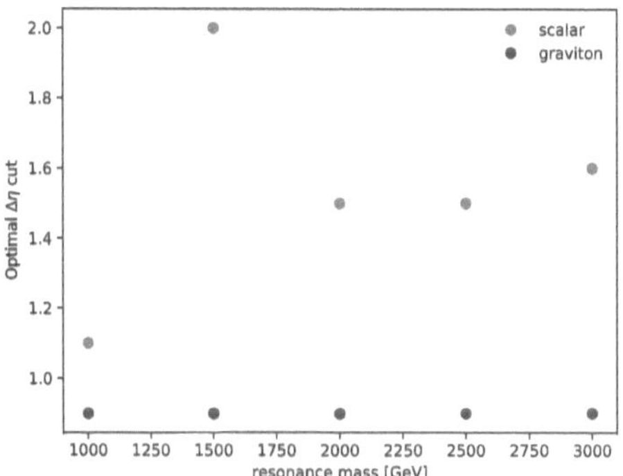

Figure 1: Optimal $|\Delta\eta|$ cut values as a function of signal mass for both spin-0 scalar and spin-2 graviton signals.

to be centered around the Higgs mass peak while expanding towards the high mass tail, where background is lower. The alternate shapes tested a few different ideas for potential improvement based on Gaussian fits to the signal $m(H_1)$ and $m(H_2)$ distributions. Figure 2 shows the mean and width of these Gaussian fits as a function of signal mass. It should be noted that the Gaussian approximation is not a particularly good one, especially for the high mass signals. Nevertheless, several alterations were designed based on these fits: moving the center of the X_{HH} distribution to $(m(H_1) = 122, m(H_2) = 112)$, making the signal region an ellipse wider in $m(H_2)$ than $m(H_1)$, making the denominator of the X_{HH} function depend on Higgs candidate p_T in addition to mass, etc. Of all the functions tested, none did more than a few percent better than the nominal X_{HH} function, as shown in Figure 3b, and so the decision was made to keep the function as-is.

A second method was also tried from the optimization of the $|\Delta\eta|$ cut. This method was to derive expected limits, without systematic uncertainties, changing the cut value of $|\Delta\eta|$ while keeping the rest of the analysis intact. An early form of the final background estimate was used for this optimization, using the older MV2 [47] b-tag algorithm at the 70% WP and without the reweighting or smoothing techniques applied. Only minor differences in the limits were observed with different cuts, an example of which is shown in Figure 4. While a tight cut value of $|\Delta\eta| < 0.9$ was observed

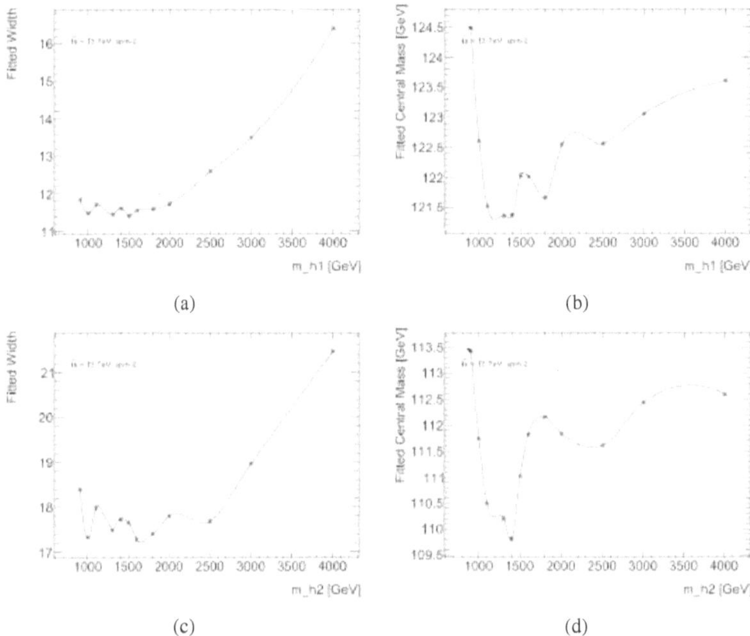

Figure 2: Mean (a, c) and width (b, d) of Gaussian distributions fit to the leading (a, b) and subleading (c, d) Higgs candidate mass distributions as a function of graviton signal mass. The Gaussian approximation breaks down at large scalar masses.

to be optimal for graviton signals, this value worsened the expected limit for high mass scalar signals. The final $|\Delta\eta|$ cut of 1.3 was chosen as a comprise, producing moderate improvement for both scalar and graviton signals. Due to the difficulty of computation, this second method was not attempted for X_{HH} optimization.

The choice of b-tagging cut was also optimized from among the DL1r 70%, 77%, and 85% working points. This optimization again done by comparing the expected limits without systematic uncertainties while keeping the rest of the analysis constant. Figure 5 shows the difference in expected limits with different b-tagging cuts. Both 77% and 85% WPs were observed to outperform the DL1r 70% WP, and all three improve the limits by approximately 10% over the MV2 70% WP used in the previous $HH \rightarrow 4b$ analysis.

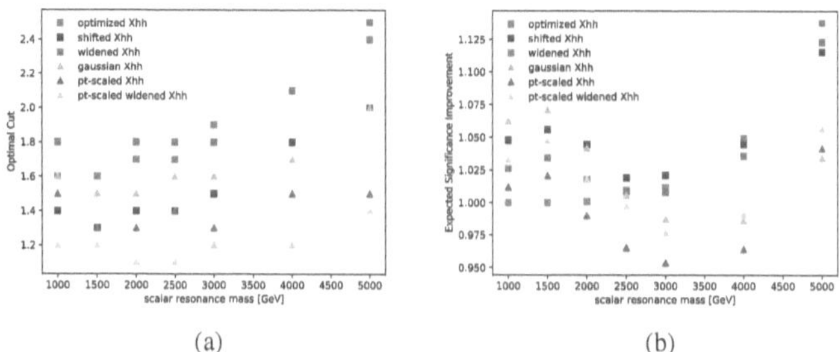

(a)   (b)

Figure 3: Optimal cut values for a set of different X_{HH} definitions as a function of scalar signal mass (a), and ratio of the expected significance of the optimal cut to the nominal $X_{HH} < 1.6$ (b). 'Optimized Xhh' refers to the nominal equation, Eq. 2.2, with optimal cut value used for each signal mass.

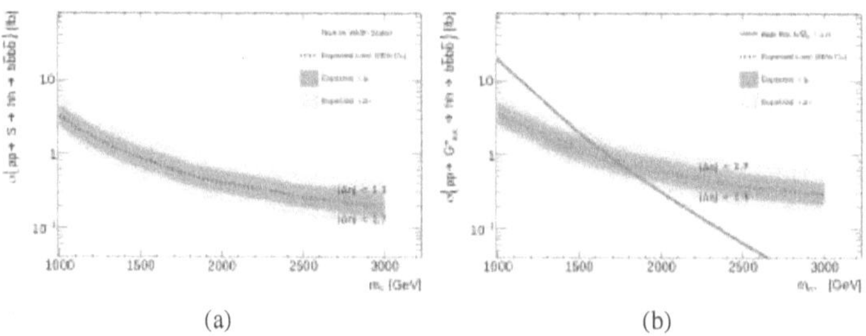

(a)   (b)

Figure 4: Comparison of expected limit with $|\Delta\eta|$ cut of 1.3 and previous value of 1.7 for (a) spin-0 and (b) spin-2 signals.

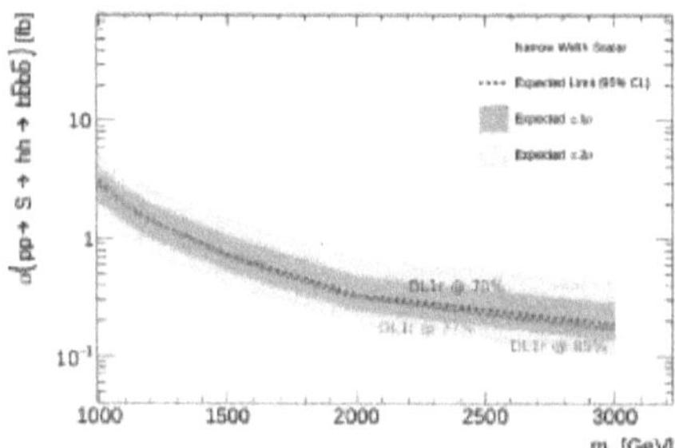

Figure 5: Comparison of expected limit for spin-0 signals with the 70%, 77%, and 85% DL1r WPs. Not shown is the approximately 10% improvement in the limits from using DL1r over MV2.

B $HH \to 4b$ **Systematics**

Most systematic uncertainties in the $HH \to 4b$ analysis are accounted for using variations of the simulation. The uncertainties on the background estimation strategy of the analysis itself are described in Section 2.5, while this appendix provides more detail on uncertainties common to many searches in ATLAS. These common variations, or the methods to derive them, are calculated from dedicated calibrations and used as Bayesian priors in the likelihood function. These priors are assumed to follow two-sided Gaussian distributions, characterized by the $\pm 1\sigma$ variations plotted here. Only those uncertainties that required additional study, beyond the default calibration, or that have an impact in the fit are discussed.

B.1 $t\bar{t}$ systematics

As mentioned in Section 2.5.4, uncertainties on many aspects of the $t\bar{t}$ simulation are considered. These include the matrix element calculation, parton shower modelling, renormalization and factorization scales, and the amount of additional hard radiation in the events.

Hard radiation The amount of hard radiation in $t\bar{t}$ decays can be adjusted through an h_{damp} parameter in the MC generator. The typical value used for ATLAS samples is $h_{\mathrm{damp}} = 1.5\, m_{\mathrm{top}}$, however samples were also generated with $h_{\mathrm{damp}} = 3\, m_{\mathrm{top}}$. Unfortunately, these additional samples contain fewer events than the nominal samples resulting in reduced precision, particularly at high masses. The alternate sample was intended to be used as the $+1\sigma$ variation, while a set of alternate showering parameters, implemented as a reweighting of the nominal sample, was intended as the -1σ variation. The reduced precision of the $+1\sigma$ sample caused difficulty when smoothing the $m(HH)$ distribution. Instead, the -1σ variation was reflected about the nominal prediction to define the uncertainty.

Final state radiation For the uncertainty in the final state radiation (FSR) α_s scale, eight sets of alternate event weights are calculated by PYTHIA 8, corresponding to μ_R parameter variations

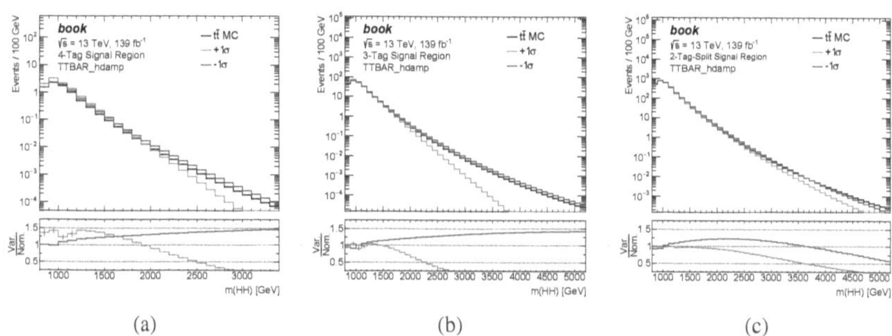

Figure 6: Parton shower uncertainty in the (a) $4b$, (b) $3b$ and (c) $2b$-split channels, derived from a comparison of PYTHIA 8 and HERWIG 7 samples. The difference between the two samples is mirrored to create symmetric uncertainty bands. The variations are correlated across channels and controlled by a single nuisance parameter in the likelihood function.

ranging between 0.5 and 2, where a value of 1 is the default. The effect of these variations on the unsmoothed distributions are shown in Figure 7 while the effect on the smoothed distributions are shown in Figure 8. While the typical uncertainty used in other ATLAS analyses has been to use the $\mu_R = 0.5$ and 2.0 variations as $\pm 1\sigma$, two issues arise in the $HH \rightarrow 4b$ case. First, the $\mu_R = 0.5$ variation shows large statistical flucations due to high weight events associated with particular phase-space of this variation. Second, all variations predict fewer events at high mass than the nominal prediction. In order to remove the potential impact of high weight events and ensure fit convergence, an uncertainty band is created by symmetrizing the $\mu_R = 2.0$ variation. This results in a smaller band than would be created by symmetrizing the $\mu_R = 0.5$ variation, for instance, but Figure 9 shows that the difference in expected limits between most conservative and the chosen option is neglible. The only mass point where a greater than 1% difference is observed is 1200 GeV, and this seems to be due to a statistical fluctuation in the $\mu_R = 0.5$ band in the $3b$ region which is then carried over to the $4b$ region as well.

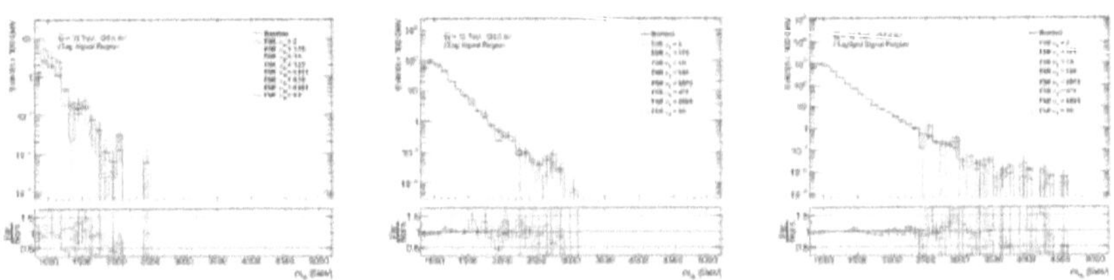

Figure 7: Uncertainty in the FSR renormalization scale before smoothing is applied in the $4b$, $3b$ and $2b$-split signal regions for a large set of variations. The $\mu_R = 2.0$ variation is symmetrized to provide an envelope for the fit.

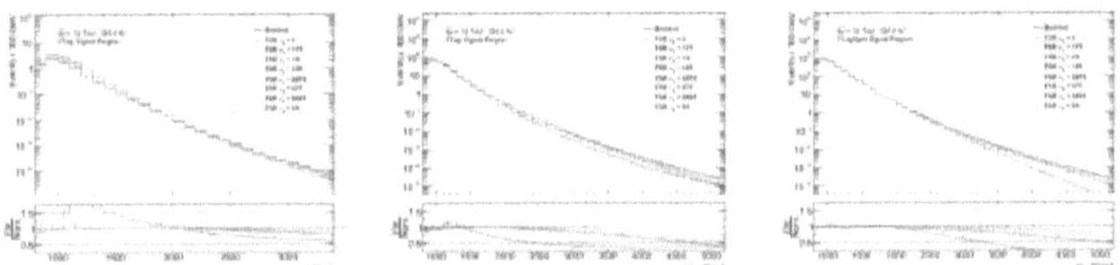

Figure 8: Uncertainty in the FSR renormalization scale after smoothing is applied in the $4b$, $3b$ and $2b$-split signal regions for a large set of variations. The $\mu_R = 2.0$ variation is symmetrized to provide an envelope for the fit.

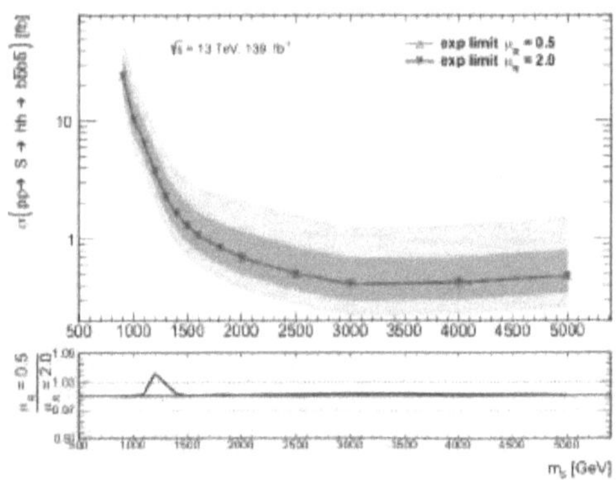

Figure 9: The difference in expected limits for the scalar signal model obtained when symmetrizing the most conservative ($\mu_R = 0.5$) variation and the chosen $\mu_R = 2.0$ variation. The difference is negligible except around 1200 GeV where a fluctuation in the $\mu_R = 0.5$ variation greatly increases the uncertainty in the $3b$ and $4b$ regions.

Parton distribution functions   The PDF uncertainty is provided as a set of 100 variations on the nominal PDF (NNPDF 2.3 LO with the A14 tune) used to generate the samples. These variations are implement as additional sets of event weights in PYTHIA 8. When evaluating the uncertainty on each plot, the full ensemble is considered and the $\pm 1\sigma$ variation in each bin corresponds to the 14th and 86th percentiles among the predictions in that bin. As shown in Figure 10, these variations are smaller than or equal to the statistical uncertainty in each bin, indicating that the uncertainty from the PDF itself is negligible. The PDF uncertainty is therefore not applied as a nuisance parameter in the final analysis.

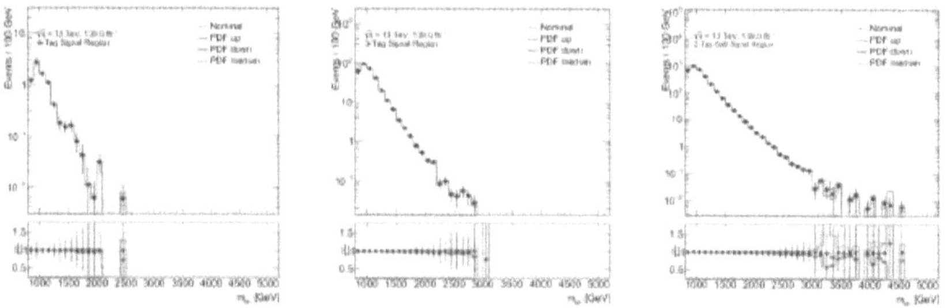

Figure 10: PDF uncertainties evaluated on the m_{HH} distribution in the $2b$-split, $3b$ and $4b$ signal regions after all cuts are applied. The up, down and median variations correspond to the 86th, 14th and 50th percentiles among the variations considered.

B.2   Jet reconstruction

The largest jet uncertainties in the $HH \to 4b$ analysis come from the mass resolution of the detector, followed by uncertainties on the modelling of the combined mass procedure. The $+1\sigma$ variation of the mass resolution is derived by smearing the Higgs and top mass peaks in MC by 20%, to match the most conservative estimate of the resolution in data. This smearing pushes a significant fraction of the signal sample out of the signal region and the effect on the total number of signal events, as function of signal mass, is shown in Figure 11. The smearing of the top mass peak pushes W jets from $t\bar{t}$ events into the signal region while pushing top jets out, resulting in the softer m_{HH} spectrum shown in Figure 12. For the resolution uncertainties, only a $+1\sigma$ variation is well-defined and the -1σ variation is created by reflecting it about the nominal prediction. The second-largest source of jet uncertainty is the modelling of the combined mass. The calibration procedure results in the bands shown in Figure 13. Many other variations of the jet modelling parameters are implemented in the $HH \rightarrow 4b$ likelihood function but have negligible impact on the result.

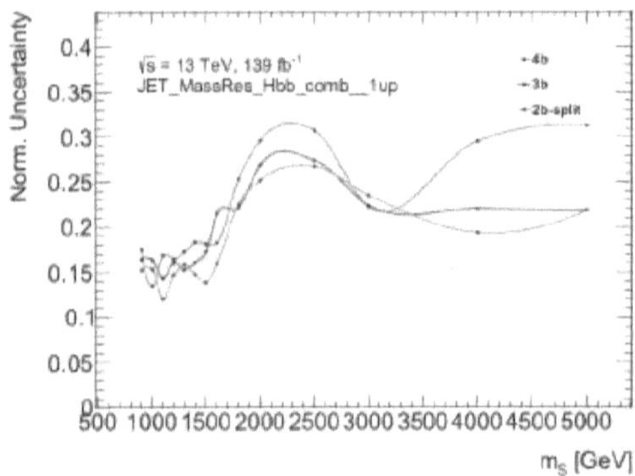

Figure 11: Normalization uncertainty, as a percentage of the spin-0 signal acceptance, due to the uncertainty on the resolution of the Higgs mass peak. A worse Higgs mass resolution worsens signal acceptance by pushing jets out of the signal region.

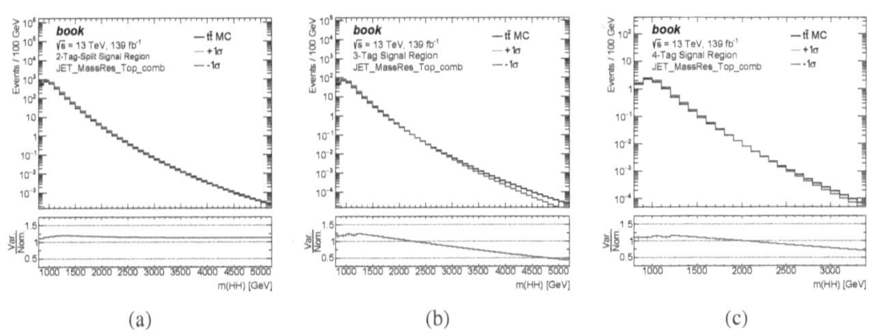

(a) (b) (c)

Figure 12: Effect of smearing the hadronic top mass peak on the $t\bar{t}$ background. The variations shown are correlated between channels and controlled by a single NP in the fit.

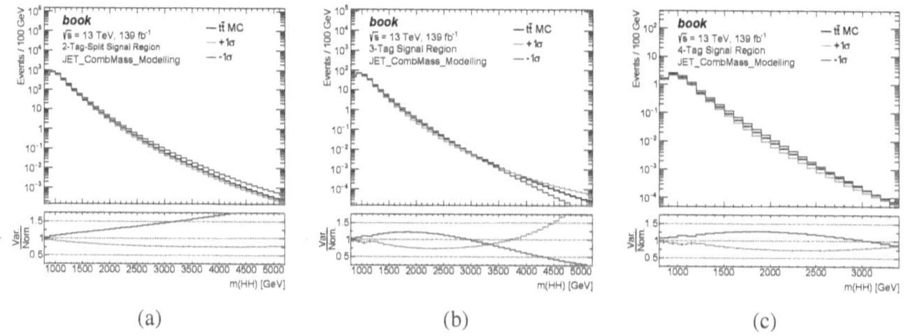

(a)   (b)   (c)

Figure 13: Large-R jet combined mass modelling uncertainty on the $t\bar{t}$ background. The variations shown are correlated between channels and controlled by a single NP in the fit.

B.3 b-tagging efficiency

The $HH \rightarrow 4b$ analysis makes extensive use of the DL1r b-tagging algorithm described in Section 1.10, the calibration of which is described in Ref. [46]. The calibration uses isolated b-jets from low-p_{T} top decays to measure the b-tagging efficiency in data and calculate scale factors. The largest uncertainty in the boosted regime comes from the high-p_{T} extrapolation uncertainty, a conservative 30% uncertainty on the scale factors assigned to jets above the highest p_{T} bin used in the calibration. The effect of that uncertainty on the $t\bar{t}$ background prediction is shown Figure 14.

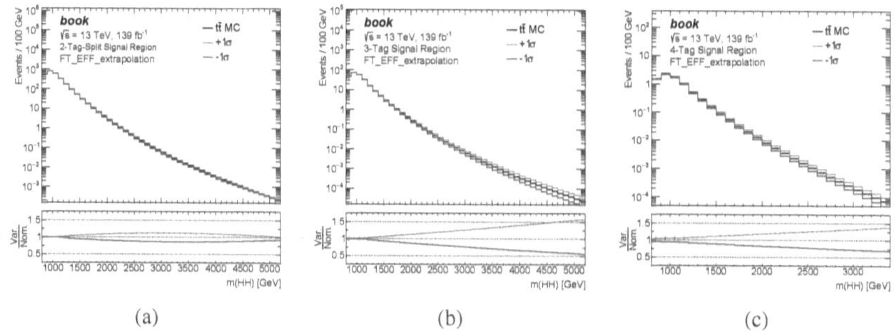

(a)   (b)   (c)

Figure 14: The effect of the high-p_{T} extrapolation uncertainty on the $t\bar{t}$ background prediction. This is the largest uncertainty on the DL1r b-tagging scale factors for jets above the p_{T} range at which the calibration was derived. The variations shown are correlated between channels and controlled by a single NP in the fit.

C   Toy Limits for the Boosted Analysis

The toy method is used to test the validity of the Wald approximation used to model the test statistic distributions in the asymptotic method. This test is particularly important at high masses where the $1/\sqrt{N}$ error term in the approximation is large. In the toy method, rather than assuming the test statistic will follow the non-central χ^2 distribution of the large N limit, the full distribution was calculated numerically using pseudo-experiments. Toys were generated under both the background-only and the signal+background hypotheses, with NPs set to their best-fit values, $\hat{\theta}(\mu)$, from a fit to data. In order to improve fit convergence, and to remove unphysical effects caused by the small negative μ phase-space, the value of the signal strength was restricted to $\mu \in [0, 10]$. The test statistic was then measured for each toy, and the resulting distributions are used to set limits.

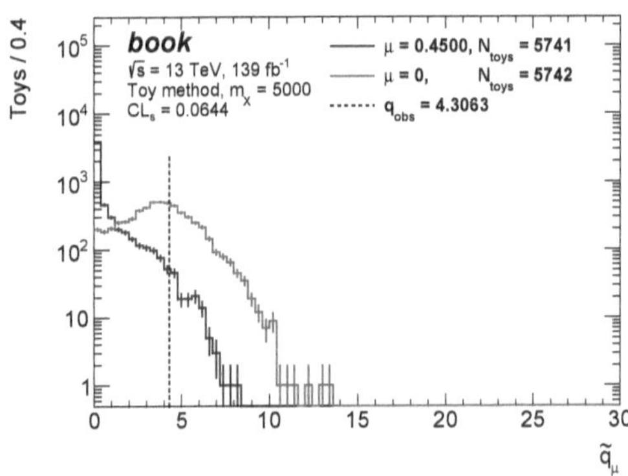

Figure 15: Test statistic distribution under both $\mu = 0$ and $\mu = 0.45$ hypotheses of the 5000 GeV spin-0 signal. The test statistic is evaluated on a set of toys randomly generated under each hypothesis.

Due to the computation time required to generate and evaluate large numbers of toys, a small initial grid of signal strengths was tested, and then additional points were added as needed to refine the estimated limits. Only masses above 1600 GeV were tested and, for the final and most precise results, only masses 3000 GeV and above were measured. The initial μ values tested were those near the asymptotic limits, and an initial set of 2000 toys were generated for each μ value. More

toys were needed for some points and up to 8000 toys were generated for some high μ values
of the 5000 GeV signals. The resulting test statistic distributions were then filtered to remove
negative values indicating failed fits before calculating p-values. An example of the test statistic
distribution for toys generated with the 5000 GeV spin-0 signal, with a strength of $\mu = 0.45$, is
shown in Figure 15.

From the test statistic distributions, p-values and CL_s are calculated as described in Section 2.6
and the 95% confidence level limit is defined by the μ value at which $\mathrm{CL}_s = 0.05$. The graph of CL_s
(μ) for the 5000 GeV spin-0 signal is shown in Figure 16, with uncertainties on the observed CL_s
due to the limited number of toys generated. To calculate the expected limit, the CL_s value for each
toy is calculated as if that toy were the observed data. The median-expected limit is then defined as
the μ value at which the median of the toy CL_s distribution is 0.05. Similarly $\pm 1\sigma$ expected limits
can be defined from the 16th and 84th quartiles of the toy CL_s distributions. Separate curves for
each quantile of the expected CL_s distribution are shown in Figure 16.

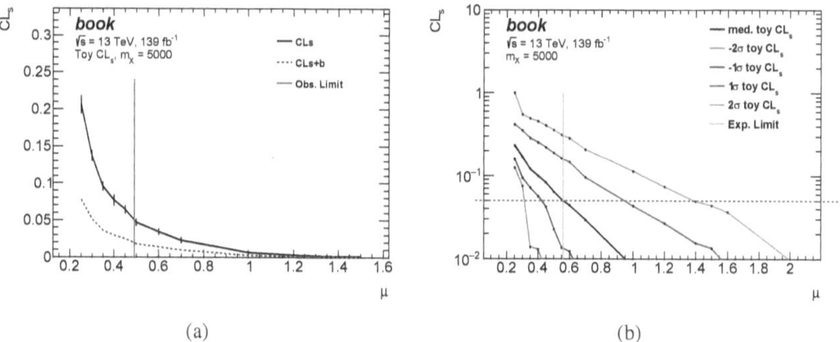

(a)   (b)

Figure 16: CL_s distributions of the 5000 GeV spin-0 signal. (a) The CL_s values of the observed
test statistic are used to calculated the observed limit, based on where the curve crosses the value
0.05. (b) Quantiles of the CL_s distributions calculated from the $\mu = 0$ toys are used to calculate the
expected limit, and the error bands on that limit.

Comparisons of the asymptotic and toy limits are shown in Figure 17, with the uncertainty band coming from the toy results. Both expected and observed limits agree with 20% between then asymptotic and toy methods, although the size of the discrepancy increases with signal mass. This discrepancy is expected due to the breakdown of the Wald approximation when the number of events per bin is small. The toy limits were used in the final results for masses above 3 TeV.

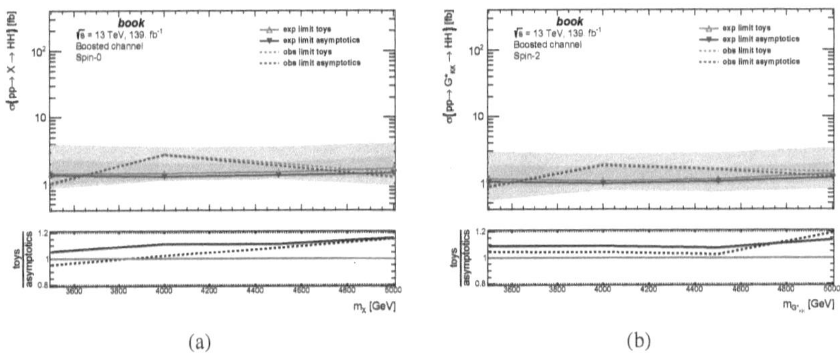

(a)   (b)

Figure 17: Comparison between asymptotic and toy limits for the (a) spin-0 and (b) spin-2 signal models. Toy limits were only calculated for masses above 1600 GeV in both models. Due to limited statistical precisions of the toys, the asymptotic limits are used in the final results.

www.ingramcontent.com/pod-product-compliance
Lightning Source LLC
LaVergne TN
LVHW051538170726
843492LV00006B/1833